Distribution Theory for Tests Based on the Sample Distribution Function

CBMS-NSF REGIONAL CONFERENCE SERIES
IN APPLIED MATHEMATICS

A series of lectures on topics of current research interest in applied mathematics under the direction of the Conference Board of the Mathematical Sciences, supported by the National Science Foundation and published by SIAM.

GARRETT BIRKHOFF, *The Numerical Solution of Elliptic Equations*

D. V. LINDLEY, *Bayesian Statistics, A Review*

R. S. VARGA, *Functional Analysis and Approximation Theory in Numerical Analysis*

R. R. BAHADUR, *Some Limit Theorems in Statistics*

PATRICK BILLINGSLEY, *Weak Convergence of Measures: Applications in Probability*

J. L. LIONS, *Some Aspects of the Optimal Control of Distributed Parameter Systems*

ROGER PENROSE, *Techniques of Differential Topology in Relativity*

HERMAN CHERNOFF, *Sequential Analysis and Optimal Design*

J. DURBIN, *Distribution Theory for Tests Based on the Sample Distribution Function*

SOL I. RUBINOW, *Mathematical Problems in the Biological Sciences*

P. D. LAX, *Hyperbolic Systems of Conservation Laws and the Mathematical Theory of Shock Waves*

I. J. SCHOENBERG, *Cardinal Spline Interpolation*

IVAN SINGER, *The Theory of Best Approximation and Functional Analysis*

WERNER C. RHEINBOLDT, *Methods of Solving Systems of Nonlinear Equations*

HANS F. WEINBERGER, *Variational Methods for Eigenvalue Approximation*

R. TYRRELL ROCKAFELLAR, *Conjugate Duality and Optimization*

SIR JAMES LIGHTHILL, *Mathematical Biofluiddynamics*

GERARD SALTON, *Theory of Indexing*

CATHLEEN S. MORAWETZ, *Notes on Time Decay and Scattering for Some Hyperbolic Problems*

F. HOPPENSTEADT, *Mathematical Theories of Populations: Demographics, Genetics and Epidemics*

RICHARD ASKEY, *Orthogonal Polynomials and Special Functions*

L. E. PAYNE, *Improperly Posed Problems in Partial Differential Equations*

S. ROSEN, *Lectures on the Measurement and Evaluation of the Performance of Computing Systems*

HERBERT B. KELLER, *Numerical Solution of Two Point Boundary Value Problems*

J. P. LASALLE, *The Stability of Dynamical Systems*

D. GOTTLIEB AND S. A. ORSZAG, *Numerical Analysis of Spectral Methods: Theory and Applications*

PETER J. HUBER, *Robust Statistical Procedures*

HERBERT SOLOMON, *Geometric Probability*

FRED S. ROBERTS, *Graph Theory and Its Applications to Problems of Society*

JURIS HARTMANIS, *Feasible Computations and Provable Complexity Properties*

ZOHAR MANNA, *Lectures on the Logic of Computer Programming*

ELLIS L. JOHNSON, *Integer Programming: Facets, Subadditivity, and Duality for Group and Semi-Group Problems*

SHMUEL WINOGRAD, *Arithmetic Complexity of Computations*

J. F. C. KINGMAN, *Mathematics of Genetic Diversity*

MORTON E. GURTIN, *Topics in Finite Elasticity*

THOMAS G. KURTZ, *Approximation of Population Processes*

JERROLD E. MARSDEN, *Lectures on Geometric Methods in Mathematical Physics*

BRADLEY EFRON, *The Jackknife, the Bootstrap, and Other Resampling Plans*

M. WOODROOFE, *Nonlinear Renewal Theory in Sequential Analysis*

D. H. SATTINGER, *Branching in the Presence of Symmetry*

R. TEMAM, *Navier–Stokes Equations and Nonlinear Functional Analysis*

J. DURBIN
London School of Economics and Political Science
University of London

Distribution Theory for Tests Based on the Sample Distribution Function

SOCIETY FOR INDUSTRIAL AND APPLIED MATHEMATICS
PHILADELPHIA

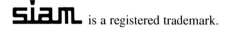 is a registered trademark.

Contents

Preface

This monograph is based on ten lectures given by me at a conference on tests based on the sample distribution function held at the State University of New York at Buffalo, in August/September 1971. These were developed from a course I gave while a Visiting Fellow at the Australian National University during the session 1970/71.

The literature on the subject is now enormous and the choice of material for inclusion reflects my predilections as a methodological statistician concerned with the development of techniques which practical workers will find useful. My main objective has been to present a coherent body of theory for the derivation of the sampling distributions of a wide range of test statistics. Efforts have been made to unify the treatment; for example, I have tried to relate the derivations for tests on the circle and the two-sample problem to the basic theory for the one-sample problem on the line. I have also placed much emphasis on the Markovian nature of the sample distribution function since this accounts for the remarkable elegance of many of the results achieved as well as the close relation with parts of the theory of stochastic processes. I have taken the opportunity to include in the sections on tests after parameters have been estimated some results obtained since the conference took place where these constitute an improvement over those presented at the time. Among omissions, the one I regret most is a discussion of the application outside the goodness-of-fit context of the techniques under study, particularly since my own interest in the area originated from the application of the Kolmogorov–Smirnov test to tests of serial correlation in time series analysis.

The written version differs from the lectures as delivered in three main respects. First, I have had to leave out some informal comments on the relation of the theory presented to practical and theoretical work in a variety of fields which I hope helped to enliven the verbal presentation. Secondly, because of the need for precision of statement in a written record, the domination of ideas by algebraical detail in the monograph is inevitably greater than in the lectures. Thirdly, I have given no indication whatever of the benefit derived by all present, and particularly by me, from the vigorous discussion by the audience of the various problems raised. Also to be recalled are the excellent talks given by other participants on their own work in the general area covered by the conference. I mention these points here in order to ensure that the dry record I have written should not obscure completely the memory of a lively social occasion.

I am extremely grateful to Professor Emanuel Parzen, Chairman of the Statistics Department at SUNYAB, for inviting me to give these lectures and for his excellent organization of the conference facilities. Thanks are due also to the Conference Board of Mathematical Sciences for administrative support and to the National Science Foundation for finance. I am indebted to Martin Knott, Colin Taylor and Roger Heard for help in removing errors from the first draft.

J. Durbin

Distribution Theory for Tests Based on the Sample Distribution Function

J. Durbin

1. Introduction.

1.1. Preliminaries. Suppose that $x_1 \leq \cdots \leq x_n$ is an ordered sample of n independent observations from a distribution with distribution function $F(x)$ and that we wish to test the null hypothesis $H_0 : F(x) = F_0(x)$. The purpose of this monograph is to study techniques for constructing and evaluating tests based on the sample distribution function $F_n(x)$ defined as the proportion of the values $x_1, \cdots, x_n \leq x$ for $-\infty \leq x \leq \infty$. Our main concern will be with the two classes of statistics typified by the Kolmogorov–Smirnov statistic

$$(1.1.1) \qquad D_n = \sup_{-\infty \leq x \leq \infty} |F_n(x) - F_0(x)|$$

and the Cramér–von Mises statistic

$$(1.1.2) \qquad W_n^2 = n \int_{-\infty}^{\infty} (F_n(x) - F_0(x))^2 \, dF_0(x),$$

respectively. Methods will be developed for finding the exact and asymptotic distributions of a wide range of variants of these statistics under both null and alternative hypotheses.

A basic result is the Glivenko–Cantelli theorem, given for continuous F_0 by Glivenko [60] and for general F_0 by Cantelli [24], which states that when $F = F_0$,

$$\Pr \left(\lim_{n \to \infty} D_n = 0 \right) = 1.$$

In consequence, tests based on D_n, and indeed most of the tests we shall consider, are strongly consistent against all alternatives, i.e., as more observations are added a false hypothesis is eventually rejected with probability one. From the standpoint of asymptotic theory this is obviously a very desirable property.

The theory to follow will be presented entirely in terms of tests of goodness of fit but this is by no means the only application of the results achieved. There are many other applications, notably to tests on the Poisson process such as

1

those considered by Cox and Lewis [29, Chap. 6], and to tests of serial independence in time series analysis as described in [43], [44].

We shall confine ourselves to the study of univariate statistics since the corresponding multivariate problems are either unsolved or are of relatively limited interest; see Pyke [95] and also [45] together with the references given therein for details of some work in the area.

It should be emphasized that from the standpoint of practical statistical analysis the role of a test of significance should be viewed as complementary to a direct examination of the data and not as a substitute for it. Thus where possible the graph of $F_n(x)$ should be plotted and inspected since this will reveal in a graphic way the nature of any substantial departure from H_0. The test of significance then provides a yardstick of comparison against which the observed discrepancies can be measured.

Useful bibliographies on tests based on the sample distribution function have been given by Darling [34] and Barton and Mallows [7]. The survey by Sahler [101] is also commended. Pyke [95] has provided a valuable review of probabilistic aspects of the subject. Stephens [119] has provided good approximations to the percentage points of many of the statistics we shall consider in an extremely compact form.

We shall use the abbreviations d.f. for distribution function and U(0, 1) for uniform distribution on the interval [0, 1]. As usual $[z]$ denotes the largest integer $\leq z$. A stochastic process will be denoted by $\{x(t)\}$ for various x and a particular sample path from the process by $x(t)$.

1.2. The sample d.f. as a stochastic process. Make the probability integral transformation $t_j = F_0(x_j)$, $j = 1, \cdots, n$, and assume that $F_0(x)$ is continuous. Then when H_0 is true, $0 \leq t_1 \leq \cdots \leq t_n \leq 1$ is an ordered sample of n independent observations from the U(0, 1) distribution. Let $F_n(t)$ be the sample d.f. for this sample, i.e., the proportion of the values $t_1, \cdots, t_n \leq t$ for $0 \leq t \leq 1$. We shall consider a number of properties of $\{F_n(t)\}$ when it is regarded as a continuous-parameter stochastic process for $0 \leq t \leq 1$. For a general introduction to the theory of stochastic processes in relation to order statistics the book by Karlin [69] can be recommended, particularly Chapter 9.

It is well known that the distribution of $t_1, \cdots, t_n$ is

$$(1.2.1) \qquad\qquad dP = n!\, dt_1 \cdots dt_n, \qquad\qquad 0 \leq t_1 \leq \cdots \leq t_n \leq 1.$$

We shall first show that this distribution can be realized as the distribution of occurrence times in a Poisson process given that n events occur in [0, 1].

Let $\{P_n(t)\}$ be the Poisson process with occurrence rate n and jumps of $1/n$ for $0 \leq t \leq 1$, i.e., $n(P_n(t_2) - P_n(t_1))$ has the Poisson distribution with mean $n(t_2 - t_1)$ for $0 \leq t_1 \leq t_2 \leq 1$, $P_n(0) = 0$ and increments are independent. Consider a set of time-points $0 < t_1 < \cdots < t_n < 1$ and choose dt_i small enough so that the intervals $[t_i, t_i + dt_i)$ are nonoverlapping. Then the probability of no event in $[0, t_1)$, one event in $[t_1, t_1 + dt_1)$, none in $[t_1 + dt_1, t_2)$, one in $[t_2, t_2 + dt_2)$, $\cdots$, none in $[t_n + dt_n, 1]$ is

$$e^{-nt_1} \cdot e^{-ndt_1} n \, dt_1 \cdot e^{-n(t_2 - t_1 - dt_1)} \cdots n \, dt_n \cdot e^{-n(1 - t_n - dt_n)} + o\left(\sum_{i=1}^{n} dt_i\right)$$

$$= n^n e^{-n} dt_1 \cdots dt_n + o\left(\sum_{i=1}^{n} dt_i\right).$$

Now the probability of n events in $[0, 1]$ is $\exp(-n)n^n/n!$. Thus the conditional probability of an event in $[t_i, t_i + dt_i)$ for $i = 1, \cdots, n$, given n events in $[0, 1]$, is

$$(1.2.2) \qquad dP = n! \, dt_1 \cdots dt_n + o\left(\sum_{i=1}^{n} dt_i\right), \qquad 0 < t_1 < \cdots < t_n < 1.$$

Letting $\max(dt_i) \to 0$ and comparing (1.2.1) and (1.2.2) we see that the two densities are the same for $0 < t_1 < \cdots < t_n < 1$. Since the events $t_i = t_j$ $(i \neq j)$, $t_1 = 0$, $t_n = 1$ in (1.2.1) have zero probability the two distributions are the same for $0 \leq t_1 \leq \cdots \leq t_n \leq 1$, i.e., the distribution of the occurrence times of $\{P_n(t)\}$ in $[0, 1]$ given $P_n(1) = 1$ is the same as that of the uniform order statistics. Since the mappings from the vector $[t_1, \cdots, t_n]'$ to the space D of functions on $[0, 1]$ which are continuous on the right with left-hand limits are the same for both $F_n(t)$ and $P_n(t)$, it follows that the distribution of the stochastic process $\{F_n(t)\}$ is the same as that of the process $\{P_n(t)\}$ given $P_n(1) = 1$. This representation of the sample d.f. as a conditioned Poisson process was used by Kolmogorov in his basic paper [74].

Many important properties of $\{F_n(t)\}$ flow from the fact that it is a Markov process. To establish this we must show that for any fixed set of values $0 < s_1 < \cdots < s_{k+1} < 1$ the conditional distribution of $F_n(s_{k+1})$ given $F_n(s_1), \cdots, F_n(s_k)$ depends only on $F_n(s_k)$. Let $r_1 = nF_n(s_1), r_j = n(F_n(s_j) - F_n(s_{j-1})), j = 2, \cdots, k + 1$. Then

$$\Pr\{nF_n(s_{k+1}) = m | nF_n(s_j), j = 1, \cdots, k\}$$

$$= \frac{n!}{r_1! \cdots r_{k+1}!(n-m)!} s_1^{r_1} \cdots (s_{k+1} - s_k)^{r_{k+1}}(1 - s_{k+1})^{n-m}$$

$$\div \frac{n!}{r_1! \cdots r_k!(n - m + r_{k+1})!} s_1^{r_1} \cdots (s_k - s_{k-1})^{r_k}(1 - s_k)^{n-m+r_{k+1}}$$

$$= \frac{(n - m + r_{k+1})!}{r_{k+1}!(n-m)!} \frac{(s_{k+1} - s_k)^{r_{k+1}}(1 - s_{k+1})^{n-m}}{(1 - s_k)^{n-m+r_{k+1}}}, \quad r_1, \cdots, r_{k+1} \geq 0,$$

$$\sum_{i=1}^{k+1} r_i = m \leq n,$$

$$= \Pr(nF_n(s_{k+1}) = m | nF_n(s_k))$$

which completes the proof. Similarly, $\{P_n(t)\}$ is a Markov process.

In order to handle distributions of Kolmogorov–Smirnov statistics a stronger property is needed. For example, a typical situation we shall meet is as follows.

Let s be the smallest value of t, if any, such that $F_n(t) = a(t)$, where $a(t)$ is a given function of t. We want to be able to say that the development of the sample path $F_n(t)$ for $t \geq s$ given a complete knowledge of the path up to time s depends only on the value at time s. It is clear that this does not follow immediately from the fact that $\{F_n(t)\}$ is a Markov process since the definition of a Markov process refers to the future development of the series given the value $F_n(s)$ at a *fixed* time s whereas in the problem under discussion s is a *random* time whose value depends on the development of the path up to time s.

The property needed to handle problems of this kind is called the strong Markov property. A full discussion requires sophisticated treatment and cannot be given here. The reader is referred to Karlin [69] for some introductory remarks and to Breiman [20] for a full treatment in relation to a variety of processes. Here, we can only outline the basic idea. Following Karlin, let s be a nonnegative random variable associated with a given continuous-parameter stochastic process $\{X(t)\}$, $0 \leq t \leq 1$. To indicate the dependence of s on the particular sample path observed, $X(t)$, we write $s = s(X(t))$. The random variable s is said to be a stopping time relative to $\{X(t)\}$ if it has the following property. If $X(\tau)$ and $Y(\tau)$ are two sample paths of the process such that $X(\tau) = Y(\tau)$ for $0 \leq \tau \leq r$ and $s(X(t)) < r$, then $s(X(t)) = s(Y(t))$. The strong Markov property is then the property that if s is a stopping time and s_0 is an arbitrarily chosen fixed time, the conditional distribution of $\{X(t)\}$, $t > s$, given $X(\tau)$, $\tau < s$, given $X(s) = x$ and given $s = s_0$, is the same as the conditional distribution of $\{X(t)\}$, $t > s_0$, given $X(s_0) = x$. This means that we can use Markovian arguments as if s were fixed even though in fact it is random, e.g., when s is the first-passage time to a boundary. Not all Markov processes have the strong Markov property but fortunately those we are concerned with possess it.

1.3. The Brownian motion processes. Our treatment of the asymptotic behaviour of the test statistics under study will be based on the properties of the *Brownian motion process* and the *tied-down Brownian motion process*. There are several ways of constructing the Brownian motion process. For the present purpose the most convenient is that of Billingsley [11, § 9]. Let C be the space of continuous functions on the interval $[0, 1]$ and let $\mathscr{C}$ be the class of Borel sets in C generated by the uniform metric

$$c(x, y) = \sup_{0 \leq t \leq 1} |x(t) - y(t)| \quad \text{for} \quad x, y \in C.$$

We ask whether there exists a stochastic process $\{w(t)\}$, $0 \leq t \leq 1$, with a probability distribution on $(C, \mathscr{C})$ possessing the properties:

    (i) $\Pr(w(0) = 0) = 1$,
    (ii) for each t in $[0, 1]$, $w(t)$ is $N(0, t)$,
    (iii) if $0 \leq t_0 \leq \cdots \leq t_k \leq 1$, then the increments $w(t_1) - w(t_0), \cdots, w(t_k) - w(t_{k-1})$ are independent.

Billingsley proves that such a process exists; we call it the Brownian motion process. Since sample paths are elements of C they are continuous. In place of

(ii) and (iii) we could have substituted the requirement that if $0 \leq t_1 \leq \cdots \leq t_k \leq 1$, then $w(t_1), \cdots, w(t_k)$ are multivariate normal with zero means and covariance function $E(w(t)w(t')) = \min(t, t')$, $0 \leq t, t' \leq 1$.

In order to study the asymptotic behaviour of $F_n(t)$ we normalize to give

$$(1.3.1) \qquad y_n(t) = \sqrt{n}(F_n(t) - t), \qquad\qquad 0 \leq t \leq 1.$$

The process $\{y_n(t)\}$ we call the *sample process*. As will be seen later, this has zero mean and covariance function

$$(1.3.2) \qquad E(y_n(t)y_n(t')) = \min(t, t') - tt', \qquad\qquad 0 \leq t, t' \leq 1.$$

For the development of asymptotic theory we wish to consider whether a normal process $\{y(t)\}$ exists in $(C, \mathscr{C})$ with zero mean and covariance function (1.3.2). It is easy to verify that if $\{w(t)\}$ is the Brownian motion process and we let

$$(1.3.3) \qquad y(t) = w(t) - tw(1), \qquad\qquad 0 \leq t \leq 1,$$

then $\{y(t)\}$ is normal with zero mean and covariance function (1.3.2). Since $\{w(t)\}$ exists in $(C, \mathscr{C})$ so does $\{y(t)\}$. Thus $\{y(t)\}$ defined by (1.3.3) has the desired properties.

For $0 < t_1 < t_2 < 1$ consider the conditional distribution of $w(t_1)$, $w(t_2)$ given $w(1) = 0$. The density of $w(t_1)$, $w(t_2)$, $w(1)$ is

$$\frac{1}{(2\pi)^{3/2}[t_1(t_2 - t_1)(1 - t_2)]^{1/2}} e^{-[w(t_1)^2/t_1 + (w(t_2) - w(t_1))^2/(t_2 - t_1) + (w(1) - w(t_2))^2/(1 - t_2)]/2}$$

and the density of $w(1)$ is $(2\pi)^{-1/2} \exp(-\tfrac{1}{2}w(1)^2)$ so the conditional joint density of $w(t_1)$ and $w(t_2)$ given $w(1) = 0$ is

$$\frac{1}{2\pi[t_1(t_2 - t_1)(1 - t_2)]^{1/2}} e^{-[t_2 w(t_1)^2/t_1(t_2 - t_1) - 2w(t_1)w(t_2)/(t_2 - t_1) + (1 - t_1)w(t_2)^2/(t_2 - t_1)(1 - t_2)]/2},$$

which is easily verified to be the joint density of two normal variables with variances $t_1(1 - t_1)$ and $t_2(1 - t_2)$ and with covariance $t_1(1 - t_2)$. Similarly, the joint density of any finite set of $w(t)$'s given $w(1) = 0$ is normal with covariance function (1.3.2). Since the distribution of a normal process in C is determined by its finite-dimensional distributions this implies that the distribution of $\{y(t)\}$ is the same as that of $\{w(t)\}$ given $w(1) = 0$. We call $\{y(t)\}$ the tied-down Brownian motion process.

Let D be the space of functions on the interval $[0, 1]$ which are right-continuous and have left-hand limits. Since the sample paths of $\{y_n(t)\}$ are elements of D we need to be able to define probability distributions on D. To this end we introduce the Skorokhod metric defined for a pair of elements $x(t)$, $y(t)$ of D by

$$(1.3.4) \qquad d(x, y) = \inf_{\lambda \in \Lambda} \left[\sup_{0 \leq t \leq 1} |x(t) - y(\lambda(t))| + \sup_{0 \leq t \leq 1} |t - \lambda(t)| \right],$$

where Λ is the class of all strictly increasing continuous functions on $[0, 1]$ such that $\lambda(0) = 0$ and $\lambda(1) = 1$ and we consider distributions on the Borel sets $\mathscr{D}$ generated by the open sets of D.

In order to study the limiting distribution of $\{y_n(t)\}$ we need to extend the domain of definition of the Brownian motion process $\{w(t)\}$ and the tied-down Brownian motion process $\{y(t)\}$ from $(C, \mathscr{C})$ to $(D, \mathscr{D})$. Following Billingsley [11, § 16], we do this by giving $\Pr(w(t) \in A)$ the value $\Pr(w(t) \in A \cap C)$ for each $A \in \mathscr{D}$, the latter probability being calculated from the distribution of $\{w(t)\}$ on $(C, \mathscr{C})$. The same is done for $\{y(t)\}$. Regarding C as a member of the class $\mathscr{D}$, we have $\Pr(w(t) \in C) = \Pr(y(t) \in C) = 1$, that is, $\{w(t)\}$ and $\{y(t)\}$, when regarded as stochastic processes in D, have continuous sample paths with probability one. Note that the definition (1.3.4) of d differs slightly from Billingsley's but this does not affect the present discussion since the two definitions are essentially equivalent.

2. Kolmogorov-Smirnov tests, finite-sample case.

2.1. Distributions of one-sided Kolmogorov-Smirnov statistics.
Let us return to the problem of testing the null hypothesis $H_0: F(x) = F_0(x)$ given an ordered sample $x_1 \leq \cdots \leq x_n$ of n independent observations from a distribution with continuous d.f. $F(x)$. We shall consider a number of tests based on the sample d.f. $F_n(x)$. A central place in our study will be occupied by Kolmogorov's [74] statistic

$$(2.1.1) \qquad D_n = \sup_{-\infty \leq x \leq \infty} |F_n(x) - F_0(x)|$$

together with the Smirnov [107], [108] one-sided statistics

$$(2.1.2) \qquad D_n^+ = \sup_{-\infty \leq x \leq \infty} [F_n(x) - F_0(x)]$$

and

$$(2.1.3) \qquad D_n^- = \sup_{-\infty \leq x \leq \infty} [F_0(x) - F_n(x)].$$

As in § 1.2 put $t_j = F_0(x_j)$, $j = 1, \cdots, n$, and let $F_n(t)$ denote the sample d.f. computed from $t_1, \cdots, t_n$ for $0 \leq t \leq 1$. Then

$$(2.1.4) \qquad \begin{aligned} D_n^+ &= \sup_{0 \leq t \leq 1} [F_n(t) - t], \\ D_n^- &= \sup_{0 \leq t \leq 1} [t - F_n(t)] \quad \text{and} \quad D_n = \max(D_n^+, D_n^-). \end{aligned}$$

Since $t_1 \leq \cdots \leq t_n$ is an ordered sample of independent observations from the $U(0, 1)$ distribution, the distributions of D_n^+, D_n^- and D_n do not depend on $F_0(x)$, i.e., they are distribution-free statistics and tests based on them are distribution-free tests. It is therefore sufficient to consider the distributions of the statistics in the form (2.1.4) where $t_1, \cdots, t_n$ come from a $U(0, 1)$ distribution. From the behaviour of $F_n(t)$ as t passes through each t_j it is easy to deduce the alternative forms

$$(2.1.5) \qquad D_n^+ = \max_{1 \leq j \leq n} \left(\frac{j}{n} - t_j \right), \qquad D_n^- = \max_{1 \leq j \leq n} \left(t_j - \frac{j-1}{n} \right).$$

For the distribution of D_n^+ we need to evaluate

$$\Pr(D_n^+ \le d) = 1 - \Pr(\text{a sample path } F_n(t) \text{ crosses the line } y = d + t),$$

where $d \ge 0$. It turns out to be just as easy to obtain the probability that a sample path crosses the general line $y = d + ct, c > 0, d \ge 0, d + c \ge 1$. For this purpose we begin by taking $d = 0$ and prove the remarkably simple result that the probability that a sample path $F_n(t)$ crosses the line $y = ct$ is $1/c$ for $c > 1$.

The path $F_n(t)$ lies wholly to the right of the line $y = ct$ if $j(nc)^{-1} < t_j$ for $j = 1, \cdots, n$, and the probability of this is

(2.1.6) $$n! \int_{c^{-1}}^{1} \int_{(n-1)(cn)^{-1}}^{t_n} \cdots \int_{2(nc)^{-1}}^{t_3} \int_{(nc)^{-1}}^{t_2} dt_1 \, dt_2 \cdots dt_n.$$

Now

$$\int_{(nc)^{-1}}^{t_2} dt_1 = t_2 - (nc)^{-1} \quad \text{and} \quad \int_{j(nc)^{-1}}^{t_{j+1}} \left[\frac{t_j^{j-1}}{(j-1)!} - \frac{t_j^{j-2}}{(j-2)!nc} \right] dt_j$$

$$= \frac{t_{j+1}^j}{j!} - \frac{t_{j+1}^{j-1}}{(j-1)!nc}$$

for $j = 2, \cdots, n$, where $t_{n+1} = 1$, so on substituting in (2.1.6) we obtain

(2.1.7) $$\Pr(F_n(t) \text{ does not cross the line } y = ct) = 1 - 1/c,$$

a result given by Daniels [32].

The result (2.1.7) provides the key to the comparative simplicity of the expression (2.1.10) below for the probability of crossing a general line. However, the proof we have given does not really explain the underlying reason why the result is true. A better insight into the problem is gained from the approach used by Vincze [125] based on the classical ballot theorem but this is too lengthy to be included here. The proof given above has been preferred because of its brevity and simplicity.

Consider now the probability that a path $F_n(t)$ crosses the general line $y = d + ct$, $0 \le d < 1, c > 0, d + c \ge 1$. Denote this line by L. Let A_j denote the point on L where $y = j/n$, i.e., the point with coordinates $[(j/n - d)/c, j/n]$ and let E_j denote the event "$F_n(t)$ crosses L at A_j but not at $A_{j+1}, \cdots, A_n$", $j = [nd + 1], \cdots, n - 1$, with E_n denoting the event "$F_n(t)$ crosses L at A_n".

The line is crossed when one of the mutually exclusive events $E_{[nd+1]}, \cdots, E_n$ occurs. Thus

(2.1.8) $$\Pr(F_n(t) \text{ crosses } L) = \sum_{j=[nd+1]}^{n} \Pr(E_j).$$

Now L is crossed at A_j when exactly j observations lie to the left of A_j and this has the binomial probability

(2.1.9) $$\binom{n}{j} \left(\frac{j/n - d}{c} \right)^j \left(\frac{c + d - j/n}{c} \right)^{n-j}.$$

Given that L is crossed at A_j, the observations $t_{j+1}, \cdots, t_n$ are distributed like an ordered uniform sample on the interval $((j/n - d)/c, 1)$. Consequently, by (2.1.7) the probability that $F_n(t)$ does not cross L at $A_{j+1}, \cdots, A_n$ given that it crosses at A_j is $1 - c_j^{-1}$, where c_j is the slope of L when the subsample is represented on the unit square, that is, $c_j = (c + d - j/n)/(1 - j/n)$ whence $1 - c_j^{-1} = (c + d - 1)/(c + d - j/n)$. Combining this with (2.1.9) we have

$$\Pr(E_j) = \frac{c + d - 1}{c}\binom{n}{j}\left(\frac{j/n - d}{c}\right)^j\left(\frac{c + d - j/n}{c}\right)^{n-j-1}$$

so on substituting in (2.1.8) we obtain finally

$$(2.1.10) \quad \Pr(F_n(t) \text{ crosses } L) = \frac{c + d - 1}{c^n} \sum_{j=[nd+1]}^{n} \binom{n}{j}\left(\frac{j}{n} - d\right)^j\left(c + d - \frac{j}{n}\right)^{n-j-1}.$$

The d.f. of D_n^+ is obtained by taking $c = 1$ in (2.1.10). This gives

$$\Pr(D_n^+ \leq d) = 1 - \Pr(F_n(t) \text{ crosses the line } y = d + t)$$

$$(2.1.11) \qquad\qquad = 1 - d \sum_{j=[nd+1]}^{n} \binom{n}{j}\left(\frac{j}{n} - d\right)^j\left(1 - \frac{j}{n} + d\right)^{n-j-1}, \qquad 0 < d < 1;$$

also $\Pr(D_n^+ \leq d) = 0$ when $d \leq 0$ and $= 1$ when $d \geq 1$. Putting $i = n - j$ we obtain the equivalent form

$$(2.1.12) \qquad \Pr(D_n^+ \leq d) = 1 - d \sum_{i=0}^{[n-nd]} \binom{n}{i}\left(\frac{i}{n} + d\right)^{i-1}\left(1 - \frac{i}{n} - d\right)^{n-i}.$$

Since by symmetry D_n^+ and D_n^- have the same distribution, (2.1.11) and (2.1.12) also give the distribution function of D_n^-. Form (2.1.11) was first given by Smirnov [108, equation (50)], and form (2.1.12) by Birnbaum and Tingey [17, equation (3.0)]. The general form (2.1.10), with i substituted for $n - j$ is due to Dempster [37]. Percentage points are given by Owen [91].

A result of Abel states that for all real a, b and integer $n \geq 0$,

$$(2.1.13) \qquad (b - n) \sum_{j=0}^{n} \binom{n}{j}(a + j)^j(b - j)^{n-j-1} = (a + b)^n,$$

where for $b = n$ the left-hand term is defined as the limit for $b \to n$. (See Birnbaum and Pyke [15, equation (5)]; a simple proof is given by Korolyuk [75]). Putting $b = n(c + d)$ and $a = -nd$ we obtain

$$(2.1.14) \qquad \frac{c + d - 1}{c^n} \sum_{j=0}^{n} \binom{n}{j}\left(\frac{j}{n} - d\right)^j\left(c + d - \frac{j}{n}\right)^{n-j-1} = 1.$$

Thus we obtain for (2.1.10) the alternative form

$$(2.1.15) \quad \Pr(F_n(t) \text{ crosses } L) = 1 - \frac{c + d - 1}{c^n} \sum_{j=0}^{[nd]} \binom{n}{j}\left(\frac{j}{n} - d\right)^j\left(c + d - \frac{j}{n}\right)^{n-j-1}$$

and for (2.1.11) the alternative form

$$(2.1.16) \qquad \Pr(D_n^+ \leq d) = d \sum_{j=0}^{[nd]} \binom{n}{j} \left(\frac{j}{n} - d\right)^j \left(1 - \frac{j}{n} + d\right)^{n-j-1}.$$

An expression similar to (2.1.15) was given by Pyke [93, equation (4)] for a special case. The general result was given by Dwass [53]. Note that while (2.1.15) seems superior to (2.1.10) for applications because it involves a shorter summation for the upper tail, it contains terms of alternating sign, whereas the terms in (2.1.10) are all positive. The effect is that (2.1.10) is generally better than (2.1.15) for computation even where it contains more terms since it is less vulnerable to rounding errors.

2.2. Distribution of the value of t at which max $(F_n(t) - t)$ occurs.

Let us call a transformation $t \rightarrow t^*$ a cyclic transformation if for t_0 fixed in $(0, 1)$, $t^* = t - t_0$ for $t_0 \leq t \leq 1$ and $t^* = 1 + t - t_0$ for $0 \leq t \leq t_0$. A set of samples $t_1 \leq \cdots \leq t_n$ which can be transformed into one another by cyclic transformations we shall call a configuration. For each fixed configuration the samples composing it may be indexed by the value of t_0 relative to a particular sample as base.

Consider the conditional distribution of t_0 given that the configuration is held fixed. Because of the uniformity of the underlying distribution this is $U(0, 1)$. Now let $\hat{t}$ be the value of t for which $F_n(t) - t$ attains its maximum. Since for each fixed configuration t_0 is $U(0, 1)$ so is $\hat{t}$. Allowing the configuration to vary it follows that the unconditional distribution of $\hat{t}$ is $U(0, 1)$. This was proved by Birnbaum and Pyke [15] by a lengthy algebraical argument. A shorter proof was given by Dwass [52]; see also Kuiper [77]. At first sight the result seems counter to intuition since one might well have expected $\hat{t}$ to have occurred relatively more frequently in the centre than at the ends of the range.

2.3. Pyke's statistics.

Pyke [93] suggested that for some purposes, instead of considering the statistic $D_n^+ = \max_j (j/n - t_j)$ it might be preferable to consider instead the modified statistic

$$(2.3.1) \qquad C_n^+ = \max_{0 \leq j \leq n} \left(\frac{j}{n+1} - t_j\right), \qquad t_0 = 0.$$

Actually Pyke took j over $1 \leq j \leq n$ but the form given seems preferable to avoid getting a negative upper terminal of summation in Pyke's formula (4). The statistics corresponding to D_n^-, D_n are

$$(2.3.2) \qquad C_n^- = \max_{0 \leq j \leq n} \left(t_j - \frac{j}{n+1}\right), \qquad C_n = \max(C_n^-, C_n^+).$$

The motivation for using the modified statistics is related to the fact that $E(t_j) = j/(n+1)$. Situations where these statistics seem more appropriate than the D_n statistics are described by Brunk [22] and in [43], [44].

Now $C_n^+ \leqq c_0$ when $j/(n + 1) - t_j \leqq c_0$ for all j, that is, when $j/n \leqq (n + 1)c_0/n$ $+ (n + 1)t_j/n$ for all j, that is, $F_n(t) \leqq (n + 1)c_0/n + (n + 1)t/n$ for all t. Using (2.1.10) and (2.1.15) we therefore have

$$\Pr(C_n^+ \leqq c_0) = 1 - \frac{(n + 1)c_0 + 1}{n + 1} \sum_{j=[(n+1)c_0+1]}^{n} \binom{n}{j}$$

(2.3.3)

$$\times \left(\frac{j}{n + 1} - c_0\right)^j \left(c_0 + 1 - \frac{j}{n + 1}\right)^{n-j-1}$$

(2.3.4)

$$= \begin{cases} \dfrac{(n + 1)c_0 + 1^{[(n+1)c_0]}}{n + 1} \displaystyle\sum_{j=0}^{} \binom{n}{j}\left(\dfrac{j}{n + 1} - c_0\right)^j \left(c_0 + 1 - \dfrac{j}{n + 1}\right)^{n-j-1} \\ \qquad\qquad\qquad\qquad\qquad\qquad\qquad\qquad \text{for} \quad 0 \leqq c_0 \leqq \dfrac{n}{n + 1}, \\ 0 \quad \text{elsewhere.} \end{cases}$$

By symmetry, the same distributions hold for C_n^-. Percentage points are given in [44].

2.4. Distributions of two-sided Kolmogorov-Smirnov statistics. Consider the distribution of Kolmogorov's statistic

$$D_n = \sup_{0 \leqq t \leqq 1} |F_n(t) - t|$$

when the underlying data are independent $U(0, 1)$ observations. Clearly,

$$\Pr(D_n \leqq d) = \Pr(D_n < d)$$

$$= \Pr(F_n(t) \text{ lies between the lines } y = \pm d + t)$$

(2.4.1) $$= \Pr(P_n(t) \text{ lies between the lines } y = \pm d + t | P_n(1) = 1)$$

$$= \frac{\Pr(P_n(t) \text{ lies between the lines } y = \pm d + t \text{ and } P_n(1) = 1)}{\Pr(P_n(1) = 1)},$$

where $\{P_n(t)\}$ is the Poisson process with occurrence rate n and jumps n^{-1} as discussed in § 1.2. We shall develop two techniques for computing the distribution function of D_n, both derived from the form (2.4.1).

Let $r = [nd]$ and $1 - \delta = nd - r$. Denote the region between the lines $y = \pm d + t$ by R and the sample path $P_n(t)$ by S. The possible points at which S can cross the line $t = j/n$ and remain inside R have ordinates $(j - r)/n, (j - r + 1)/n, \cdots, (j + r)/n$. Denote these points by $A_{j1}, \cdots, A_{jp}$, where $p = 2r + 1$.

Given that S passes through A_{ji}, the probability that it passes through $A_{j+1,k}$ is the probability of exactly $k - i + 1$ observations in the interval $(j/n, (j + 1)/n]$, that is, $e^{-1}/(k - i - 1)!$, for $i, k = 2, 3, \cdots, p - 1$ and $k \geqq i - 1$. In moving from A_{j1} to $A_{j+1,k}$, $k = 1, \cdots, p - 1$, the path S will remain in R only if k observations occur in $(j/n, (j + 1)/n)$ at least one of which occurs in the interval $(j/n, (j + 1 - \delta)/n)$. The probability of this is $e^{-1}(1 - \delta^k)/k!$. Similarly, the probability of S moving from A_{ji} to $A_{j+1,p}$ and remaining in R is $e^{-1}(1 - \delta^{p-i+1})/(p - i - 1)!$. Finally, the probability of moving from A_{j1} to $A_{j+1,p}$ in R is $e^{-1}(1 - 2\delta^p)/p!$ for $\delta \leq \frac{1}{2}$ and is $e^{-1}(1 - 2\delta^p + (2\delta - 1)^p)/p!$ for $\delta \geq \frac{1}{2}$.

Let $e^{-j}u_{ji}$ be the probability that S passes through A_{ji} while remaining inside R and let $u'_j = [u_{j1}, \cdots, u_{jp}]$, $j = 0, 1, \cdots, n$. The transition from u_j to u_{j+1} is given by the relation

$$(2.4.2) \qquad u_{j+1} = Hu_j, \qquad j = 0, 1, \cdots, n - 1,$$

where the transition matrix is given by

$$(2.4.3) \quad H = \begin{bmatrix} 1 - \delta & 1 & 0 & 0 & \cdots & 0 \\ \dfrac{1 - \delta^2}{2!} & 1 & 1 & 0 & & \vdots \\ \dfrac{1 - \delta^3}{3!} & \dfrac{1}{2!} & 1 & 1 & & 0 \\ \vdots & \vdots & & & & 1 \\ \dfrac{1 - 2\delta^p + (2\delta - 1)^p_+}{p!} & \dfrac{1 - \delta^{p-1}}{(p-1)!} & \cdots & \dfrac{1 - \delta^2}{2!} & & 1 - \delta \end{bmatrix},$$

v. here $z_+ = z$ if $z > 0$ and $= 0$ if $z \leq 0$. The initial conditions are specified by the vector u_0 which has unity in the $[nd + 1]$th position and zeros elsewhere. Applying (2.4.2) repeatedly we obtain $u_n = H^n u_0$. The probability that S passes through $(1,1)$ and remains in R is e^{-n} times the $[nd + 1]$th element of this. Now the unconditional probability that S passes through $(1, 1)$ is simply the probability of n observations in $(0, 1)$, that is, $e^{-n}n^n/n!$. Using (2.4.1) we therefore obtain finally

$$(2.4.4) \qquad \begin{aligned} \Pr(D_n \leq d) &= \frac{\Pr(S \text{ passes through } (1, 1) \text{ and remains in } R)}{\Pr(S \text{ passes through } (1, 1))} \\ &= \frac{n!}{n^n} \times \text{ the } ([nd + 1], [nd + 1])\text{th element of } H^n). \end{aligned}$$

This derivation is based on the treatment in [42] of the more general problem of calculating the probability that $F_n(t)$ remains between a pair of parallel lines $ny = a + (n + c)t$ and $ny = -b + (n + c)t$ (we shall return to this shortly). The relations (2.4.2) and (2.4.4) are generalizations of expressions [16] and [17] given for the case nd an integer in Kolmogorov's [74] paper; for this case they were used by Birnbaum [12] to tabulate the distribution function of D_n.

Let

$$(2.4.5) \qquad q_n(d) = \frac{n^n}{n!} \Pr(D_n \leq d) \quad \text{for} \quad n = 1, 2, \cdots \text{ with } q_0(d) = 1$$

and let $f(z) = \sum_{n=0}^{\infty} q_n(d)z^n$. Kemperman [70, equation (5.40)], showed that

$$(2.4.6) \qquad f(z) = \frac{g(z, nd)^2}{g(z, 2nd)},$$

where

$$g(z, b) = \sum_{j=0}^{[b]} \frac{(j - b)^j}{j!} z^j.$$

This implies the difference equation

(2.4.7) $$\sum_{j=0}^{[2nd]} \frac{(j - 2nd)^j}{j!} q_{r-j}(d) = 0, \quad r = 2[nd] + 1, 2[nd] + 2, \cdots,$$

with initial conditions

$q_0(d) = 1,$

(2.4.8) $$q_r(d) = \begin{cases} \dfrac{r^r}{r!} & \text{for} \quad r = 1, \cdots, [nd], \\[2ex] \dfrac{r^r}{r!} - \dfrac{2nd}{r!} \sum_{j=0}^{[r-nd]} \binom{r}{j} (j + nd)^{j-1} (r - j - nd)^{r-j} \end{cases}$$

$$\text{for} \quad r = [nd] + 1, \cdots, 2[nd].$$

The difference equation (2.4.7) was given by Massey [85] for the case nd an integer. We have included this discussion of it because of its intrinsic interest; for practical computation (2.4.4) is preferable since it is less vulnerable to rounding errors than (2.4.7), the terms of which have alternating signs.

Actually, Kemperman's form of the generating function is more general than (2.4.6) since it gives the probability that $F_n(t)$ lies between a pair of lines $y = d_1 + t$ and $y = -d_2 + t$ with d_1, d_2 positive but not necessarily equal.

Now we consider the extension of (2.4.2) and (2.4.4) to give the probability $p_n(a, b, c)$ that $F_n(t)$ lies between the lines $ny = a + (n + c)t$, $ny = -b + (n + c)t$, where c is a positive integer and $a, b, a + c, b - c > 0$. We need this extension to enable us to compute the distribution of the two-sided Pyke statistic C_n and Kuiper's [78] statistic $V_n = D_n^+ + D_n^-$ which we shall consider in § 5.2. By an argument which is basically the same as that given above for the special case $a = b = nd$ and $c = 0$, it is shown on p. 402 of [42] that corresponding to (2.4.3) we have

(2.4.9) $$p_n(a, b, c) = \frac{n!}{(n + c)^n} \times \text{the } ([b - c + 1], [b + 1])\text{th element of } H_1^{n+c},$$

where

(2.4.10) $$H_1 = \begin{bmatrix} 1 - \delta & 1 & 0 & \cdots\cdots\cdots & 0 \\[2ex] \dfrac{1 - \delta^2}{2!} & 1 & 1 & & \\[2ex] \dfrac{1 - \delta^3}{3!} & \dfrac{1}{2!} & 1 & & 0 \\[1ex] \vdots & \vdots & & & 1 \\[2ex] \dfrac{1 - \delta^p - \varepsilon^p + h}{p!} & \dfrac{1 - \varepsilon^{p-1}}{(p-1)!} & \cdots & \dfrac{1 - \varepsilon^2}{2!} & 1 - \varepsilon \end{bmatrix},$$

in which $p = [b - c] + [a + c] + 1$, $\delta = [b - c] + 1 - (b - c)$, $\varepsilon = [a + c] + 1$
$- (a + c)$ and $h = 0$ if $\delta + \varepsilon \leq 1$ and $h = (\delta + \varepsilon - 1)^p$ if $\delta + \varepsilon > 1$.

Putting $q_n(a, b, c) = (n + c)^n p_n(a, b, c)/n!$, $n = 1, 2, \cdots$ with $q_0 = 1$ and letting

(2.4.11)
$$f(z) = \sum_{r=0}^{\infty} q_r(a, b, c) z^{r - [c]}$$

we obtain

$$f(z) = \frac{g(z, a) g(z, b - c)}{g(z, a + b)},$$

where g is as before. This result, given in [42] for the case c a positive integer was
shown by Steck [111] to be valid generally. The expression (2.4.11) leads immediately to the difference equation

(2.4.12)
$$\sum_{j=0}^{[a+b]} \frac{(j - a - b)^j}{j!} q_{r-j}(a, b, c) = 0,$$

$$r = -[c] + [a + c] + [b - c] + 1, \ -[c] + [a + c] + [b - c] + 2, \cdots$$

with $q_s(a, b, c) = 0$ for $s < 0$. Initial conditions can be deduced from (2.4.11).

Inequalities for the two-sided probabilities in terms of one-sided probabilities
have been given in [42].

2.5. The probability that $F_n(t)$ crosses a general boundary.
For a variety of
reasons we need to be able to compute the probability that $F_n(t)$ crosses a general
boundary, but apart from exceptional cases such as those discussed by Whittle
[136], explicit formulae are only available for the straight-line case. However, it is
an easy matter to develop straightforward computing algorithms by exploiting
the Markovian properties of $F_n(t)$ as we shall show below.

To illustrate the need for a treatment of the general-boundary case, consider
the problem of computing the power of the D_n^+ test of the null hypothesis H_0 that
a set of observations come from a distribution with continuous d.f. $F_0(x)$ against
the alternative hypothesis H_1 that the (continuous) d.f. is $F_1(x)$.

Let $t = F_0(x)$, $t' = F_1(x)$ and let $F_n(t)$, $F_n'(t')$ be the sample d.f.'s relative to t, t'
respectively. Then $F_n'(t')$ has the same distribution when H_1 is true as does $F_n(t)$
when H_0 is true. Now H_0 is rejected when $F_n(t) \geq d + t$, that is, $F_n'(t') \geq d$
$+ F_0(F_1^{-1}(t'))$. The rejection probability on H_1 is thus the same as the probability
that $F_n(t)$ crosses the boundary $d + F_0(F_1^{-1}(t))$ when H_0 is true. Similarly, the power
of the D_n test is the probability that $F_n(t)$ crosses either boundary $d \pm F_0(F_1^{-1}(t))$
when H_0 is true.

A further application is to enable us to compute the significance points and
powers of generalized Kolmogorov–Smirnov statistics of the form

$$K_n^+ = \sup_{0 \leq t \leq 1} [\psi_1(t)(F_n(t) - t)],$$

(2.5.1)
$$K_n^- = \sup_{0 \leq t \leq 1} [\psi_2(t)(t - F_n(t))],$$

$$K_n = \max (K_n^+, K_n^-),$$

where ψ_1, ψ_2 are nonnegative weight functions intended to give greater weight to discrepancies at some parts of the $(0, 1)$ interval than at others. Note that in general $\psi_1(t), \psi_2(t)$ may be zero for some values of t. For example, the Renyi statistic

$$(2.5.2) \qquad \sup_{a \leq F(x) \leq b} \frac{F_n(x) - F(x)}{F(x)}$$

belongs to this class (see Renyi [98] and Birnbaum and Lientz [14]). The statistic K_n, with $\psi_1 = \psi_2$, was proposed by Anderson and Darling [4].

Since $K_n^+ > d$ when $F_n(t) > t + d/\psi_1(t)$, where $d/\psi_1(t)$ is taken as infinite when $\psi_1(t) = 0$, $\Pr(K_n^+ > d)$ is obtained as the probability that $F_n(t)$ crosses the boundary $t + d/\psi_1(t)$. Similarly, the power of the test based on K_n^+ of H_0 against H_1 is the probability that $F_n(t)$ crosses the boundary $F_0(F_1^{-1}(t)) + d/\psi_1[F_0(F_1^{-1}(t))]$, with corresponding results for K_n^- and K_n.

Since $F_n(t)$ can only take values j/n for $j = 0, 1, \cdots, n$, all problems whose boundaries have only a finite number of points with ordinates j/n can be put in the following form. Let A_j be the point with coordinates $(a_j, r_j/n), j = 1, \cdots, m$, where $0 \leq a_1 \leq \cdots \leq a_m \leq 1$, $r_j \in (0, 1, \cdots, n)$ and where m is the number of points on the boundary with ordinates of the form j/n. We require the probability that $F_n(t)$ passes through at least one of the points $A_1, \cdots, A_m$. We compute this as the probability that the Poisson process $\{P_n(t)\}$ passes through at least one of $A_1, \cdots, A_m$ given $P_n(1) = 1$.

Let α_j' be the probability that $P_n(t)$ passes through A_j but not through $A_1, \cdots, A_{j-1}, j = 1, \cdots, m$. The event "$P_n(t)$ passes through A_k" is the union of the mutually exclusive events "$P_n(t)$ passes through A_j but not $A_1, \cdots, A_{j-1}$ and also passes through A_k", $j = 1, \cdots, k$. Since the probability that $P_n(t)$ passes through A_k is $\exp(-na_k)(na_k)^{r_k}/r_k!$ and the probability of $P_n(t)$ passing through A_k given that it has passed through A_j is $\exp[-n(a_k - a_j)][n(a_k - a_j)]^{r_k - r_j}/(r_k - r_j)!$ for $r_k \geq r_j$ and is zero otherwise, we obtain

$$(2.5.3) \qquad \frac{e^{-na_k}(na_k)^{r_k}}{r_k!} = \alpha_k' + \sum_j^k \frac{e^{-n(a_k - a_j)}[n(a_k - a_j)]^{r_k - r_j}}{(r_k - r_j)!} \alpha_j',$$

where $\sum_j^k$ indicates summation over all $j < k$ such that $r_j \leq r_k$. Let $\alpha_j = e^{na_j} n^{-r_j} \alpha_j'$. Substituting in (2.5.3) we obtain

$$(2.5.4) \qquad \frac{a_k^{r_k}}{r_k!} = \alpha_k + \sum_j^k \frac{(a_k - a_j)^{r_k - r_j}}{(r_k - r_j)!} \alpha', \qquad k = 1, \cdots, m,$$

which is solved recursively for $\alpha_1, \cdots, \alpha_m$.

The probability that $P_n(t)$ passes through at least one A_j and $P_n(1) = 1$ is

$$\sum_{j=1}^m \frac{e^{-n(1 - a_j)}[n(1 - a_j)]^{n - r_j}}{(n - r_j)!} \alpha_j' = e^{-n} n^n \sum_{j=1}^m \frac{(1 - a_j)^{n - r_j}}{(n - r_j)!} \alpha_j.$$

But $\Pr(P_n(1) = 1) = e^{-n}n^n/n!$. Hence,

$$\Pr(F_n(t) \text{ passes through at least one of } A_1, \cdots, A_m)$$

(2.5.5)
$$= n! \sum_{j=1}^{m} \frac{(1 - a_j)^{n-r_j}}{(n - r_j)!} \alpha_j,$$

where $\alpha_1, \cdots, \alpha_m$ are the solutions of (2.5.4).

Let us apply this method to the case of a single boundary defined by the strictly increasing continuous function $y = a(t)$ where $a(0) > 0$ and $a(1) > 1$. Define a_j by the relations $a(a_j) = j/n, j = p, p + 1, \cdots, n$, where $p = [na(0)] + 1$. From (2.5.4) and (2.5.5) we deduce that the probability that $F_n(t)$ crosses the boundary is

(2.5.6)
$$n! \sum_{j=p}^{n} \frac{(1 - a_j)^{n-j}}{(n - j)!} \alpha_j,$$

where $\alpha_p, \cdots, \alpha_n$ are the solutions of the recursion

(2.5.7)
$$\frac{a_k^k}{k!} = \alpha_k + \sum_{j=p}^{k-1} \frac{(a_k - a_j)^{k-j}}{(k - j)!} \alpha_j, \qquad k = p, p + 1, \cdots, n,$$

the final term being taken to be zero for $k = p$. The corresponding solution for a pair of boundaries is given in (36), (37) and (40) of [47] on which the above treatment is based.

Many different forms of the recursion for the single-boundary case considered above have been given by Wald and Wolfowitz [127], Daniels [32], Noé and Vandewiele [90], Suzuki [122] and Knott [73]. For the two-boundary case an extremely interesting recursion, different from the above, has been derived by Epanechnikov [54] and Steck [111]. These authors express the probability that $F_n(t)$ lies wholly between the two boundaries as

(2.5.8)
$$\Pr(u_i \leq t_i \leq v_i, i = 1, \cdots, n) = n! Q_n,$$

where $0 \leq t_1 \leq \cdots \leq t_n \leq 1$ is the ordered $U(0, 1)$ sample, u_i and v_i are given numbers satisfying $0 \leq u_i \leq v_i \leq 1$ for $i = 1, \cdots, n$, and where Q_n satisfies the recursion

(2.5.9)
$$Q_k = (v_k - u_k)_+ Q_{k-1} - \frac{(v_{k-1} - u_k)_+^2}{2!} Q_{k-2} + \cdots + \frac{(-1)^{k-1}(v_1 - u_k)_+^k}{k!},$$
$$k = 1, 2, \cdots, n,$$

in which $z_+ = z$ if $z > 0$ and $= 0$ if $z \leq 0$. As shown by Steck, relations (2.5.9) are equivalent to the explicit determinantal form

(2.5.10)
$$Q_k = \begin{vmatrix} (v_1 - u_1)_+ & \dfrac{(v_1 - u_2)_+^2}{2!} & \cdots & \dfrac{(v_1 - u_k)_+^k}{k!} \\ 1 & (v_2 - u_2)_+ & \cdots & \dfrac{(v_2 - u_k)_+^{k-1}}{(k-1)!} \\ 0 & 1 & \ddots & \vdots \\ \vdots & & \ddots & \vdots \\ 0 \cdots\cdots\cdots 0 & & 1 & (v_k - u_k)_+ \end{vmatrix}.$$

A proof of this result will be given incidentally in § 7.2 while deriving the characteristic function of an estimator based on paths lying between the two boundaries.

The recursion technique is easily adapted to deal with grouped data. Suppose, for example, that there are g groups with probability contents $p_1, \cdots, p_g$, where $\sum_1^g p_i = 1$, and that the observed number in the ith group is n_i, where $\sum_1^g n_i = n$. Let $s_j = \sum_1^j p_i$, $s_0 = 0$ and $\tilde{F}_n(t) = n^{-1} \sum_1^{j-1} n_i$, $s_{j-1} \leqq t < s_j$, $j = 1, \cdots, g$. Then $E(\tilde{F}_n(s_j)) = s_j$. Let

$$(2.5.11) \qquad\qquad D_{nk}^+ = \max_{j=1,\cdots,g-1} [\tilde{F}_n(s_j) - s_j].$$

Then $D_{nk}^+ > d$ if and only if $\tilde{F}_n(t)$ crosses the boundary $y = d + t \equiv L$, say.

Now the event "$\tilde{F}_n(t)$ crosses L when $s_{k-1} \leqq t < s_k$" is the union of the disjoint events "$\tilde{F}_n(t)$ crosses L for the first time when $s_{j-1} \leqq t < s_j$ and subsequently crosses again when $s_{k-1} \leqq t < s_k$", $j = 1, \cdots, k$. Introducing the Poisson process $\{P_n(t)\}$ as before, on taking probabilities we obtain the recursion

$$(2.5.12) \qquad\qquad a_k = \alpha_k + \sum_{j=1}^{k-1} \alpha_j a_{jk}, \qquad\qquad k = 2, 3, \cdots,$$

with $\alpha_1 = a_1$, where

$$a_j = \Pr [s_{j-1} + d \leq P_n(s_j) < \min (s_j + d, 1)], \qquad\qquad j = 1, 2, \cdots,$$

$$a_{jk} = \Pr [s_{k-1} + d \leq P_n(s_k) < \min (s_k + d, 1)|s_{j-1} + d \leq P_n(s_j) < \min (s_j + d, 1)],$$

$$j = 1, \cdots, k - 1; k = 2, 3, \cdots,$$

$$\alpha_j = \Pr [s_{i-1} + d \leq P_n(s_i) < \min (s_i + d, 1) \text{ for the first time when } i = j],$$

$$j = 1, 2, \cdots.$$

Now $\Pr (P_n(1) = 1) = n^n e^{-n}/n!$. Thus

$$\Pr (\tilde{F}_n(t) \text{ crosses } L) = n^{-n} n! e^n \sum_{j=1}^g \alpha_j b_j,$$

where

$$b_j = \Pr [P_n(1) = 1|s_{j-1} + d \leq P_n(s_j) < \min (s_j + d, 1)].$$

One has to substitute the appropriate expressions for the Poisson probabilities in the above, though as before the exponential factors drop out and can be ignored. The method can be extended to two boundaries and to curved boundaries without any difficulty of principle, though of course the formulae become more complicated.

Schmid [102] has considered the related problem of the distributions of D_n^+ and D_n for discontinuous distributions.

3. Asymptotic theory of Kolmogorov-Smirnov tests.

3.1. Limiting finite-dimensional distributions of the sample process. As before, let $F_n(t)$ denote the sample d.f. for a sample of n independent $U(0, 1)$ observations. From the fact that $[nF_n(t), nF_n(t') - nF_n(t), n - nF_n(t')]$ is a trinomial variable, we deduce

$$E(F_n(t)) = t,$$

$$C(F_n(t), F_n(t')) = n^{-1}t(1 - t'), \qquad 0 \le t \le t' \le 1.$$

Let

(3.1.1) $$y_n(t) = \sqrt{n}(F_n(t) - t)$$

and remove the restriction $t \le t'$. We obtain

(3.1.2) $$E(y_n(t)) = 0,$$

(3.1.3) $$C(y_n(t), y_n(t')) = \min(t, t') - tt', \qquad 0 \le t, t' \le 1.$$

From the application of the central limit theorem to the multinomial distribution we deduce immediately that the vector $[y_n(s_1), \cdots, y_n(s_k)]'$ is asymptotically normally distributed with mean zero and covariance matrix determined by (3.1.3) for all $0 \le s_1, \cdots, s_k \le 1$. But (3.1.2) and (3.1.3) are the mean and covariance functions of the tied-down Brownian motion process $\{y(t)\}$ considered in §1.3. It follows that the finite-dimensional distributions of the process $\{y_n(t)\}$ converge to those of $\{y(t)\}$ as $n \to \infty$.

3.2. Weak convergence of the sample d.f. It is natural to conjecture that a function of $y_n(t)$ such as $\sqrt{n} D_n = \sup_t |y_n(t)|$ will converge in distribution to the corresponding function of $y(t)$, namely $\sup_t |y(t)|$. This idea is due to Doob [38] and provides a powerful technique for finding limiting distributions of suitable functions of $y_n(t)$. However, the convergence of the finite-dimensional distributions is not in itself sufficient to validate the conjecture. A complete treatment of other conditions that must be satisfied requires the technical apparatus of the theory of weak convergence, a detailed exposition of which can be found in Billingsley [11]. There is only space here to give the salient points of the theory as it applies to the sample d.f.

The sequence of stochastic processes $\{y_n(t)\}$, $n = 1, 2, \cdots$, in $(D, \mathcal{D})$ is said to *converge weakly* or to *converge in distribution* to the process $\{y(t)\}$ and we write $\{y_n(t)\} \xrightarrow{\mathcal{D}} \{y(t)\}$ if $\lim_{n \to \infty} E[f(y_n(t))] = E[f(y(t))]$ for all real bounded functions f on D which are continuous in metric d. That this holds for the convergence of the sample d.f. to tied-down Brownian motion is proved in [11, Chap. 3].

The practical value of weak convergence theory arises from the fact that if g is a measurable function on D which is continuous almost everywhere in metric d with respect to the distribution of $\{y(t)\}$ and if $\{y_n(t)\} \xrightarrow{\mathcal{D}} \{y(t)\}$, then $g(y_n(t))$ converges in distribution to $g(y(t))$ (see Billingsley [11, Theorem 5.1]). Sufficient conditions are that g is continuous in d at all points of D and $\{y_n(t)\} \overset{\mathscr{D}}{\to} \{y(t)\}$. Thus to prove that $\sqrt{n}\, D_n \overset{\mathscr{D}}{\to} \sup_t |y(t)|$ it is sufficient to show that the function $g(x(t)) = \sup_t |x(t)|$ is continuous in d for all $x(t) \in D$.

Now if $d(x, x') < \varepsilon$, there is a $\lambda \in \Lambda$ such that

$$\sup_{0 \le t \le 1} |x(t) - x'\{\lambda(t)\}| + \sup_{0 \le t \le 1} |t - \lambda(t)| < 2\varepsilon$$

so that

$$\sup_{0 \le t \le 1} |x(t) - x'\{\lambda(t)\}| < 2\varepsilon.$$

Now

$$\sup_{0 \le t \le 1} |x(t)| \le \sup_{0 \le t \le 1} |x'\{\lambda(t)\}| + \sup_{0 \le t \le 1} |x(t) - x'\{\lambda(t)\}|$$

$$\le \sup_{0 \le t \le 1} |x'(t)| + 2\varepsilon.$$

Similarly one can show that $\sup_t |x(t)| \ge \sup_t |x'(t)| - 2\varepsilon$, whence $\sup_t |x(t)|$ is continuous in metric d. It follows that $\sqrt{n}\, D_n$ is distributed asymptotically as $\sup_t |y(t)|$, where $\{y(t)\}$ is tied-down Brownian motion. In the same way D_n^+ and D_n^- are distributed asymptotically as $\sup_t (y(t))$.

Similar results hold for the K_n^+ statistic (2.5.1) whenever $\sqrt{n}\, K_n^+ = \sup_t \psi_1(t) y_n(t)$ is continuous in d. This does not apply to the Anderson–Darling statistic with $\psi_1(t) = [t(1 - t)]^{-1/2}$ since continuity does not hold in this case. Indeed the asymptotic distribution of this statistic is degenerate. Similar remarks apply to K_n^- and K_n.

An alternative approach to the weak convergence of $\{y_n(t)\}$ has been given by Breiman [20, Chap. 13], and independently by Brillinger [21]. Essentially, this uses a Skorokhod construction to obtain a sequence of processes $\{y_n'(t)\}$ and a process $\{y'(t)\}$ having the same distribution as $\{y_n(t)\}$ and $\{y(t)\}$ respectively, such that with probability one $y_n'(t) \to y'(t)$ uniformly as $n \to \infty$. This implies that $\{y_n'(t)\} \overset{\mathscr{D}}{\to} \{y'(t)\}$ and hence that $\{y_n(t)\} \overset{\mathscr{D}}{\to} \{y(t)\}$. Although somewhat indirect, this approach is mathematically simpler than Billingsley's and has the advantage of yielding explicit error bounds as Brillinger shows. However, Billingsley's treatment is preferred here since his exposition in [11] is lucid and comprehensive and is presented together with a great deal of relevant ancillary material.

3.3. Distribution of $\sup_t (y(t))$.

Since sample paths of $\{y(t)\}$ in D are continuous with probability one we may restrict our attention to continuous paths. Let $\delta^+ = \sup_t (y(t))$, let S' denote a sample path of $\{y(t)\}$ and let L denote the line $y = d, d > 0$. Then $\Pr(\delta^+ > d) = \Pr(S' \text{ crosses } L)$.

We now exploit the fact demonstrated in § 1.3 that the distribution of $\{y(t)\}$ is the same as that of $\{w(t)\}$ given $w(1) = 0$, where $\{w(t)\}$ is Brownian motion. Let S denote a sample path of $\{w(t)\}$. We have

$$\Pr(\delta^+ > d) = \Pr(S \text{ crosses } L | w(1) = 0)$$

(3.3.1)

$$= \lim_{\partial w \to 0} \frac{\Pr(S \text{ crosses } L \text{ and } 0 \le w(1) \le \partial w)}{\Pr(0 \le w(1) \le \partial w)}.$$

Now

$$\Pr (S \text{ crosses } L \text{ and } 0 \leq w(1) \leq \partial w) = \int_0^1 f(s)g(s) \, ds \, \partial w + o(\partial w)$$

(3.3.2)

$$= \int_0^1 f(s) \frac{e^{-d^2/2(1-s)}}{[2\pi(1-s)]^{1/2}} ds \, \partial w + o(\partial w),$$

where $f(s)$ is the first-passage density of paths S to L at $t = s$ and where $g(s)$ is the density of $w(1)$ at $w(1) = 0$ given $w(s) = d$, i.e., $g(s) = [2\pi(1 - s)]^{-1/2} \exp[-\frac{1}{2}d^2/(1 - s)]$. The truth of (3.3.2) follows from the strong Markov property of $\{w(t)\}$.

Similarly,

$$\Pr (S \text{ crosses } L \text{ and } 2d \leq w(1) \leq 2d + \partial w) = \int_0^1 f(s)g^*(s) \, ds \, \partial w + o(\partial w)$$

(3.3.3)

$$= \int_0^1 f(s) \frac{e^{-d^2/2(1-s)}}{[2\pi(1-s)]^{1/2}} ds \, \partial w + o(\partial w),$$

where $g^*(s)$ is the density of $w(1)$ at $w(1) = 2d$ given that $w(s) = d$, that is, $g^*(s) = [2\pi(1 - s)]^{-1/2} \exp[-\frac{1}{2}d^2/(1 - s)]$. But since every path satisfying $2d \leq w(1) \leq 2d + \partial w$ must cross L, (3.3.3) must equal the unconditional probability that $2d \leq w(1) \leq 2d + \partial w$, that is, $(2\pi)^{-1/2} \exp[-\frac{1}{2}(2d)^2] \, \partial w + o(\partial w)$. Since (3.3.2) and (3.3.3) are identical we deduce that

(3.3.4) $$\Pr (S \text{ crosses } L \text{ and } 0 \leq w(1) \leq \partial w) = \frac{e^{-2d^2}}{\sqrt{(2\pi)}} \partial w + o(\partial w).$$

Also, $\Pr (0 \leq w(1) \leq \partial w) = (2\pi)^{-1/2} \partial w + o(\partial w)$. Substituting in (3.3.1) we therefore obtain

$$\Pr (\delta^+ > d) = \lim_{\partial w \to 0} \frac{(2\pi)^{-1/2} e^{-2d^2} \partial w + o(\partial w)}{(2\pi)^{-1/2} \partial w + o(\partial w)}$$

(3.3.5)

$$= e^{-2d^2}.$$

This result was originally obtained by Smirnov [108] as the limit of $\Pr (\sqrt{n} D_n^+ > d)$ as $n \to \infty$ (see (3.6.1) below). Note that although the method of proof used above resembles the classical reflection principle it is in fact simpler since we have avoided the problem of defining what is meant by the concept of "equi-probable paths".

By considering in place of $\{w(t)\}$ the process $\{w(t) + ct\}$ one can show by the same method that Pr (a sample path of $\{y(t)\}$ crosses the line $y = d + ct$) $= \exp[-2d(d + c)]$ when $d, d + c > 0$. This result was obtained by Doob [38] by using the fact that if $\{y(t)\}$ is tied-down Brownian motion and $w(t) = (1 + t) \cdot y[t/(1 + t)]$, then $\{w(t)\}$ is Brownian motion. Doob then uses a reflection technique which is different from and lengthier than the above.

3.4. Distribution of sup, $|y(t)|$. Let $\delta = \sup_t |y(t)|$, let S and S' be as in § 3.3 and let $L(a)$ denote the line $y = a + t, 0 \leq t \leq 1$. Then

$$\Pr(\delta > d) = \Pr(S' \text{ crosses } L(d) \text{ or } L(-d))$$

$$(3.4.1) \qquad = \Pr(S \text{ crosses } L(d) \text{ or } L(-d)|w(1) = 0)$$

$$= \lim_{\partial w \to 0} \frac{\Pr(S \text{ crosses } L(d) \text{ or } L(-d) \text{ and } 0 \leq w(1) \leq \partial w)}{\Pr(0 \leq w(1) \leq \partial w)}.$$

Let A_p be the event "S crosses $L(d)$ then $L(-d)$ then $L(d) \cdots$ (p crossings) and $0 \leq w(1) \leq \partial w$", $p = 1, 2, \cdots$, and let B_q be the event "S crosses $L(-d)$ then $L(d)$ then $L(-d) \cdots$ (q crossings) and $0 \leq w(1) \leq \partial w$", $q = 1, 2, \cdots$. On allowing for double-counting of paths by the method of inclusion and exclusion, for example, as in Feller [55, p. 99], we obtain

$$(3.4.2) \qquad \begin{aligned} &\Pr(S \text{ crosses } L(d) \text{ or } L(-d) \text{ and } 0 \leq w(1) \leq \partial w) \\ &= \sum_{j=1}^{\infty} (-1)^{j-1}(\Pr(A_j) + \Pr(B_j)). \end{aligned}$$

(See Blackman [18] for details of a similar calculation for the two-sample problem.) From (3.3.4),

$$(3.4.3) \qquad \Pr(A_1) = \Pr(B_1) = \frac{e^{-2d^2}}{\sqrt{(2\pi)}} \partial w + o(\partial w).$$

Let $f(s, a)$ be the first-passage density of paths of $\{w(t)\}$ to the line $y = a$ at $t = s$ and let $g(s, a)$ be the conditional density of $w(1)$ at $w(1) = 0$ given $w(s) = a, 0 < s < 1$. By considering paths which first meet $L(d)$ at time s, which subsequently first meet $L(-d)$ at time t and which satisfy $0 \leq w(1) \leq \partial w$, we deduce that

$$(3.4.4) \qquad \Pr(A_2) = \int_0^1 \int_s^1 f(s, d)f(t - s, -2d)g(t, -d) \, dt \, ds \, \partial w + o(\partial w)$$

on using the strong Markov property of $\{w(t)\}$. On making the transformation $s' = t - s, t' = t$ in (3.4.4) we obtain

$$(3.4.5) \quad \Pr(A_2) = \int_0^1 \int_{s'}^1 f(s', -2d)f(t' - s', d)g(t', -d) \, dt' \, ds' \, \partial w + o(\partial w).$$

But the right-hand side of (3.4.5) is the probability of a path which first meets $L(-2d)$ at time s', which subsequently first meets $L(-d)$ at time t' and which satisfies $0 \leq w(1) \leq \partial w$ and this is simply the probability of a path which meets $L(-2d)$ and satisfies $0 \leq w(1) \leq \partial w$, that is,

$$(3.4.6) \qquad \frac{e^{-2(2d)^2}}{\sqrt{(2\pi)}} \partial w + o(\partial w)$$

from (3.3.4), since every path which meets $L(-2d)$ must subsequently meet $L(-d)$ if it also satisfies $0 \leq w(1) \leq \partial w$. By symmetry the same expression (3.4.6) is obtained for $\Pr(B_2)$.

A heuristic way of understanding this result is to say that what the transformation $(s, t) \to (s', t')$ does essentially is to construct a new sample path S^*, equiprobable with S, by interchanging the increments in $w(t)$ from the intervals $(0, s)$ and (s, t). Thus, whenever S crosses $L(d)$ then $L(-d)$, S^* crosses $L(-2d)$. The difficulty with this argument is that the concept of equiprobable paths is hard to make precise.

The same technique can be used for calculating $\Pr(A_j) = \Pr(B_j)$ for any j. Consider, for example, a path S in B_{2m} which first meets $L(-d)$ at time s_1, subsequently first meets $L(d)$ at time s_2, then meets $L(-d)$ again at time s_3 and so on and finally satisfies $0 \leq w(1) \leq \partial w$. Arguing heuristically, we construct an equiprobable path S^*, which meets $L(md)$, by interchanging increments in $w(t)$ from intervals $(0, s_1)$ and (s_{2m-1}, s_{2m}), from intervals (s_2, s_3) and (s_{2m-3}, s_{2m-2}), and so on. We deduce that $\Pr(B_{2m})$ is just the probability that a path meets $L(2md)$ and satisfies $0 \leq w(1) \leq \partial w$, that is,

$$(3.4.7) \qquad \Pr(B_{2m}) = \Pr(A_{2m}) = \frac{e^{-2(2m)^2 d^2}}{\sqrt{(2\pi)}}, \qquad m = 1, 2, \cdots.$$

Alternatively, arguing precisely, we have

$$\Pr(B_{2m}) = \int_0^1 \int_0^{s_{2m}} \cdots \int_0^{s_3} \int_0^{s_2} f(s_1, -d)f(s_2 - s_1, 2d)f(s_3 - s_2, -2d) \cdots$$
$$f(s_{2m} - s_{2m-1}, 2d)g(s_{2m}, d)\, ds_1 \cdots ds_{2m}$$
$$= \int_0^1 I(s_{2m})g(s_{2m}, d)\, ds_{2m},$$

where

$$I(s_{2m}) = \int_0^{s_{2m}} \cdots \int_0^{s_2} f(s_1, -d)f(s_2 - s_1, 2d) \cdots f(s_{2m} - s_{2m-1}, 2d)\, ds_1 \cdots ds_{2m-1}$$

$$= \int \cdots \int_{\substack{r_1, \cdots r_{2m-1} \geq 0 \\ \sum_1^{2m-1} r_j = s_{2m}}} f(r_1, -d)f(r_2, 2d) \cdots f(r_{2m-1}, 2d)\, dr_1 \cdots dr_{2m-1}$$

$$= \int \cdots \int_{\substack{r_1', \cdots r_{2m-1}' \geq 0 \\ \sum_1^{2m-1} r_j' = s_{2m}}} f(r_1', 2d)f(r_2', 2d) \cdots f(r_{2m-1}', -2d)f(r_{2m}', -d)\, dr_1' \cdots dr_{2m-1}',$$

where we have put $r_i = s_i - s_{i-1}(s_0 = 0)$ and $r_{2i-1}' = r_{2m-2i+1}$, $r_{2i}' = r_{2i}$, $i = 1, \cdots, m$. Putting $s_j' = \sum_1^j r_i'$ and substituting for $I(s_{2m})$ we obtain

$$\Pr(B_{2m}) = \int_0^1 \int_0^{s_{2m}} \cdots \int_0^{s_3'} \int_0^{s_2'} f(s_1', 2d)f(s_2' - s_1', 2d) \cdots$$
$$f(s_{2m} - s_{2m-1}', -d)g(s_{2m}, d)\, ds_1' \cdots ds_{2m-1}'\, ds_{2m},$$

which is simply the probability that a path meets $L(2md)$ and satisfies $0 \leq w(1)$ $\leq \partial w$. Using (3.4.6) we obtain (3.4.7).

In the same way we obtain

$$\Pr(A_{2m+1}) = \Pr(B_{2m+1}) = \frac{e^{-2(2m+1)^2 d^2}}{\sqrt{(2\pi)}}, \qquad m = 1, 2, \cdots,$$

so that on substitution in (3.4.2) we obtain

$$\Pr(S \text{ crosses } L(d) \text{ or } L(-d) \text{ and } 0 \leq w(1) \leq \partial w) = 2 \sum_{j=1}^{\infty} (-1)^{j-1} \frac{e^{-2j^2 d^2}}{\sqrt{(2\pi)}} \partial w$$

$$+ o(\partial w).$$

Now $\Pr(0 \leq w(1) \leq \partial w) = (2\pi)^{-1/2} \partial w + o(\partial w)$. Hence on substituting in (3.4.1) we obtain

$$\Pr(\delta > d) = 2 \sum_{j=1}^{\infty} (-1)^{j-1} e^{-2j^2 d^2},$$

whence

(3.4.8) $$\Pr(\delta \leq d) = \sum_{j=-\infty}^{\infty} (-1)^j e^{-2j^2 d^2} = \lim_{n \to \infty} \Pr(\sqrt{n} D_n \leq d).$$

This is the basic result of Kolmogorov's classic paper [74]. Kolmogorov obtained (3.4.8) as the solution of a limiting diffusion equation satisfying appropriate boundary conditions. A simpler derivation was given by Doob [38] which was based, like the above, on the evaluation of the probability that a sample path of $\{w(t)\}$ crosses one or both of two straight-line boundaries. However, the details of Doob's derivation are more complicated.

The form (3.4.8) converges very rapidly when d is large and is therefore very convenient for tests based on the upper tail of the distribution, which is the usual situation. If one wishes to compute a probability in the lower tail the alternative form

(3.4.9) $$\Pr(\delta \leq d) = \frac{\sqrt{(2\pi)}}{d} \sum_{j=1}^{\infty} e^{-(2j-1)^2 \pi^2 / 8d^2}$$

converges more rapidly. The fact that (3.4.8) and (3.4.9) are equivalent follows from a standard relation in theta functions (van der Pol and Bremmer [124, p. 236, equation (11)]).

By a straightforward extension of the above method one can show that the probability that a sample path of $\{w(t)\}$ lies entirely in the region between the two parallel lines $y = a + ct$ and $y = -b + ct$ where $c, b, a + c, b - c > 0$ is

(3.4.10) $$1 - \sum_{j=1}^{\infty} \{ e^{-2[ja+(j-1)b][ja+(j-1)b+c]} + e^{-2[(j-1)a+jb][(j-1)a+jb-c]}$$

$$- e^{-2j(a+b)(ja+jb+c)} - e^{-2j(a+b)(ja+jb-c)} \}.$$

3.5. The probability that $y(t)$ crosses a general boundary. In order to be able to compute asymptotic powers of D_n^+, D_n^- and D_n and asymptotic significance points and powers of the generalized statistics K_n^+, K_n^- and K_n of (2.5.1) we need to be able to calculate the probability that a sample path of $\{y(t)\}$ crosses a general boundary or one or both of a pair of such boundaries. Let us take first the case of a single boundary a defined by $y = a(t)$. As before we evaluate this probability as the probability that a sample path of $\{w(t)\}$ crosses a given $w(1) = 0$. Following the treatment in [47] we assume that $a(t)$ has a bounded derivative for $0 \leq t \leq 1$. Let $\alpha(s)$ be the probability that a sample path of $\{w(t)\}$ crosses a before time s, that is, $w(t) > a(t)$ for some t in $0 < t < s$. The probability that for t fixed $a(t) \leq w(t) \leq a(t) + dw$ is the sum over disjoint subintervals $[s_j, s_{j+1})$ covering $[0, t)$ of the probabilities of the mutually exclusive events "a sample path crosses a for the first time in $[s_j, s_{j+1})$ and $a(t) \leq w(t) \leq a(t) + dw$". Now

$$\Pr(a(t) \leq w(t) \leq a(t) + dw) = (2\pi t)^{-1/2} \exp[-\tfrac{1}{2}a(t)^2/t]\,dw + o(dw)$$

and

$$\Pr(a(t) \leq w(t) \leq a(t) + dw | w(s) = a(s))$$
$$= [2\pi(t - s)]^{-1/2} \exp[-\tfrac{1}{2}(a(t) - a(s))^2/(t - s)]\,dw + o(dw), \qquad s < t,$$

Dividing by dw and letting $\max(s_{j+1} - s_j) \to 0$, $dw \to 0$, we therefore obtain the integral equation

$$(3.5.1) \qquad \frac{1}{\sqrt{(2\pi t)}} e^{-a(t)^2/2t} = \int_0^t \frac{1}{[2\pi(t - s)]^{1/2}} e^{-(a(t) - a(s))^2/2(t - s)}\,d\alpha(s), \qquad 0 \leq t \leq 1,$$

for the unknown $\alpha(t)$. Integral equations of this type were given by Fortet [58] for a general class of diffusion processes.

The probability that $y(t)$ first crosses a in $(s, s + ds)$ is the same as the conditional probability of the event "$w(t)$ first crosses a in $(s, s + ds)$" given $w(1) = 0$. Since the conditional density of $w(1)$ at $w(1) = 0$ given $w(s) = a(s)$ is $[2\pi(1 - s)]^{-1/2} \exp[-\tfrac{1}{2}a(s)^2/(1 - s)]$ and since the unconditional density of $w(1)$ at $w(1) = 0$ is $(2\pi)^{-1/2}$, the required conditional probability is $(1 - s)^{-1/2} \exp[-\tfrac{1}{2}a(s)^2/(1 - s)]\,d\alpha(s)$. Integrating with respect to s gives for the probability that a path of $\{y(t)\}$ crosses a,

$$(3.5.2) \qquad P_a = \int_0^1 (1 - s)^{-1/2} \exp[-\tfrac{1}{2}a(s)^2/(1 - s)]\,d\alpha(s).$$

Since explicit solutions can only be obtained in very special cases, for most problems occurring in practice other than the straight-line case, (3.5.1) has to be solved and (3.5.2) evaluated by numerical methods. Of course once $\alpha(s)$ is known the numerical evaluation of (3.5.2) is a routine task so we concentrate our attention on the solution of (3.5.1). This is a Volterra equation of the first kind and the difficulty in solving it arises from the presence of the term $(t - s)^{-1/2}$ in the kernel since this $\to \infty$ as $s \to t$. Three methods have been suggested for dealing with this.

The first is based essentially on the replacement of (3.5.1) by the system of linear recursions

(3.5.3)
$$\frac{1}{\sqrt{(2\pi t)}} e^{-a(t)^2/2t} = \sum_{i=0}^{t/h-1} g(ih, t)\delta\alpha(ih), \qquad t = h, 2h, \cdots, 1,$$

to be solved for $\delta\alpha(ih) = \alpha(\overline{i+1}\,h) - \alpha(ih)$, where $g(ih, t)$ is the conditional density of $w(t)$ at $w(t) = a(t)$, given that $w(s)$ has crossed a for the first time in the interval $ih < s \leq \overline{i+1}\,h$. The length of subinterval h is chosen so that h^{-1} is an integer.

It should be noted that no approximation is involved at this stage—the relations (3.5.3) are exact. The problem is to find a method of estimating $g(ih, t)$ which is sufficiently accurate as well as computationally feasible. The treatment in [47] proceeds by exploiting known results for crossing probabilities for straight-line boundaries. Two techniques are suggested. In the first, the assumption is made that the first-passage density $\alpha'(s)$ is constant over the interval $(ih, \overline{i+1}\,h)$ and in the second it is assumed that $\alpha'(s)$ is proportional to the first-passage density of paths of $\{w(t)\}$ to a straight-line segment L_i approximating $a(t)$ over the interval $(ih, \overline{i+1}\,h)$. In both cases the appropriate integrations are carried out along L_i for $ih < s \leq \overline{i+1}\,h$. Numerical illustrations are given which show that the method can work very effectively for sufficiently smooth $a(t)$.

An alternative approach, due to Weiss and Anderssen [133], is based on the recognition that the singularity in (3.5.1) is due only to the term $(t - s)^{-1/2}$. The remaining part of the kernel, namely, $(2\pi)^{-1/2} \exp[-\frac{1}{2}(a(t) - a(s))^2/(t - s)]$ can be expected to vary fairly slowly. Assuming that this is sufficiently well approximated by its mid-point value $(2\pi)^{-1/2} \exp[-\frac{1}{2}(a(t) - a(\overline{i + \frac{1}{2}}\,h))^2/(t - \overline{i + \frac{1}{2}}\,h)] = \phi(ih, t)$, say, and that $\alpha'(s)$ is constant within subintervals we obtain as our approximation to (3.5.1),

(3.5.4)
$$\frac{1}{\sqrt{(2\pi t)}} e^{-a(t)^2/2t} = \sum_{i=0}^{t/h-1} \phi(ih, t) \int_{ih}^{\overline{i+1}\,h} \frac{ds}{\sqrt{(t - s)}} \delta\alpha(ih)$$

$$= 2 \sum_{i=0}^{t/h-1} \phi(ih, t)([t - ih]^{1/2} - [t - \overline{i+1}\,h]^{1/2}) d\alpha(ih)$$

$$t = h, 2h, \cdots, 1.$$

These equations are then solved recursively. Further developments of the approach are given by Anderssen, de Hoog and Weiss [6]. Having obtained $\delta\alpha(ih)$ one may then estimate p_a in (3.5.2) by

(3.5.5)
$$\sum_{i=0}^{1/h-1} (1 - \overline{i + \frac{1}{2}}\,h) \exp[-\frac{1}{2}a(\overline{i + \frac{1}{2}}\,h)^2/(1 - \overline{i + \frac{1}{2}}\,h)] d\alpha(ih).$$

A third technique for solving (3.5.1) based on a standard method of solving Abel integral equations, to which (3.5.1) is closely related, has been given by Smith [109].

Extensions to the two-boundary case are given in [47] and [6]; the former reference also illustrates the way in which the solution may be used to calculate the asymptotic power of the Kolmogorov–Smirnov test.

3.6. Series expansions for the distribution functions of Kolmogorov-Smirnov statistics. In an immediate category between exact expressions for finite samples and limiting distributions are the results of Smirnov [108] who showed that

$$(3.6.1) \qquad \Pr(\sqrt{n}D_n^+ \leq d) = 1 - e^{-2d^2}\left(1 - \frac{2d}{3\sqrt{n}} + O\left(\frac{1}{n}\right)\right),$$

and Lauwerier [80] who obtained a complete series expansion for this probability in inverse powers of $\sqrt{n}$. For the two-sided case, let

$$\Phi(a,b) = \lim_{n \to \infty} \Pr(\sqrt{n}D_n^+ \leq a, \sqrt{n}D_n^- \leq b)$$

$$(3.6.2)$$

$$= 1 + \sum_{j=1}^{\infty} \{2 e^{-2j^2(a+b)^2} - e^{-2[ja+(j-1)b]^2} - e^{-2[(j-1)a+jb]^2}\}$$

from (3.4.10). Darling [35] obtained the expression

$$\Pr(\sqrt{n}D_n^+ \leq a, \sqrt{n}D_n^- \leq b) = \Phi(a,b) + \frac{1}{6\sqrt{n}}\left(\frac{\partial\Phi}{\partial a} + \frac{\partial\Phi}{\partial b}\right) + O\left(\frac{1}{n}\right).$$

Further results are given by Li-Chien Chang [26], [27] and Korolyuk [75].

3.7. Powers of Kolmogorov-Smirnov tests. Techniques for computing the exact and asymptotic powers of Kolmogorov–Smirnov tests have been given in §§ 2.5 and 3.5. Without going into details we mention here the main references to other work on this topic.

Massey [85], [86], [87] gave a lower bound to the power of the test based on the two-sided statistic D_n against general continuous alternatives. He also showed that the test is biased. Birnbaum [13] obtained exact upper and lower bounds for the one-sided statistic D_n^- and the asymptotic form for the lower bound, again against general continuous alternatives. Chapman [28] gave related results for more restricted classes of alternatives and made comparisons with other test statistics. Further asymptotic upper and lower bounds are given by Quade [96].

Capon [25] considered the limiting power of the two-sample two-sided statistic D_{nm} against parametric families of alternatives relative to the optimum likelihood-ratio test. He obtained a lower bound to an analogue of the limiting Pitman efficiency as the probabilities of both Type I and Type II errors tend to zero in an appropriate fashion and gave explicit results for particular cases; for example, for alternatives consisting of a shift in mean of a normal distribution he obtained the limit $2/\pi = 0\cdot637$ and for a shift in scale of a normal distribution he obtained the limit $(\pi e)^{-1} = 0\cdot117$. In contrast, Klotz [72] considered limiting power from the standpoint of Bahadur efficiency, i.e., keeping the alternative fixed and letting the Type I error probability tend to zero at an appropriate rate as sample size increases. For the statistic D_n^+ he obtained expressions for Bahadur efficiency which, for the normal location and scale alternatives, converge to the above values $2/\pi$ and $(\pi e)^{-1}$ as the alternatives approach the null hypothesis. Further results from the standpoint of Bahadur efficiency, both for the generalized Kolmogorov–Smirnov statistic $\sup_x \sqrt{n} |F_n(x) - F(x)| \psi(F(x))$ and for the Kuiper statistic V_n, were given by Abrahamson [1]. Hoadley [64] gave some related material. Hájek and Sidák [63, §§ VI 4.4–4.6] have given expressions for the limiting powers of general classes of statistics of Kolmogorov–Smirnov types.

A device based on re-ordering of intervals between successive t_j''s was suggested in [41] which leads to an improvement in power in some situations, for example, when testing for exponentiality as demonstrated by Seshadri, Csörgo and Stephens [103].

4. Cramér–von Mises tests.

4.1. The test statistics. Let us return to the basic problem of testing the hypothesis that a sample of n independent observations come from a distribution with specified continuous d.f. $F_0(x)$. Cramér [30, p. 145], proposed as a measure of the discrepancy between the sample d.f. $F_n(x)$ and $F_0(x)$ the statistic

$$J_n = \int_{-\infty}^{\infty} (F_n(x) - F_0(x))^2 \, dx$$

and this was generalized by von Mises [126] to the form

$$\omega_n^2 = \int_{-\infty}^{\infty} g(x)(F_n(x) - F_0(x))^2 \, dx,$$

where $g(x)$ is a suitably chosen weight function. Smirnov [105], [106] modified this to

$$(4.1.1) \qquad W_n^2 = n \int_{-\infty}^{\infty} \psi(F_0(x))(F_n(x) - F_0(x))^2 \, dF_0(x)$$

so as to give a distribution-free statistic. Putting $t = F_0(x)$ as before we obtain

$$(4.1.2) \qquad W_n^2 = n \int_0^1 \psi(t)(F_n(t) - t)^2 \, dt.$$

Statistics of the form (4.1.1), together with related forms such as (4.1.9) and (5.4.1) below we shall refer to collectively as statistics of Cramér–von Mises type. We shall also follow common usage and call the special case of (4.1.1) with $\psi = 1$ the Cramér–von Mises statistic and write it as

$$(4.1.3) \qquad W_n^2 = n \int_{-\infty}^{\infty} (F_n(x) - F_0(x))^2 \, dF_0(x) = n \int_0^1 (F_n(t) - t)^2 \, dt.$$

The special case of (4.1.1) with $\psi(F_0(x)) = [F_0(x)(1 - F_0(x))]^{-1}$ we shall call the Anderson–Darling statistic after the authors who first studied it in detail in [4] and we shall write it as

$$(4.1.4) \qquad A_n^2 = n \int_{-\infty}^{\infty} \frac{(F_n(x) - F_0(x))^2}{F_0(x)(1 - F_0(x))} \, dF_0(x) = n \int_0^1 \frac{(F_n(t) - t)^2}{t(1 - t)} \, dt.$$

This form is of particular interest in view of the fact that $E[n(F_n(t) - t)^2] = t(1 - t)$, i.e., discrepancies are weighted by the reciprocals of their standard deviations.

Putting $F_n(t) = j/n$ for $t_j \leq t < t_{j+1}$ and integrating we obtain for (4.1.2) the alternative form

$$(4.1.5) \qquad W_n^2 = 2 \sum_{j=1}^{n} \left[\phi_2(t_j) - \frac{j - \frac{1}{2}}{n} \phi_1(t_j) \right] + n \int_0^1 (1 - t)^2 \psi(t)\, dt,$$

where

$$(4.1.6) \qquad \phi_1(t) = \int_0^t \psi(s)\, ds, \qquad \phi_2(t) = \int_0^t s\,\psi(s)\, ds$$

whenever these integrals exist, while for (4.1.3) and (4.1.4) we obtain

$$(4.1.7) \qquad W_n^2 = \sum_{j=1}^{n} \left(t_j - \frac{j - \frac{1}{2}}{n} \right)^2 + \frac{1}{12n},$$

and

$$(4.1.8) \qquad A_n^2 = -n - \frac{1}{n} \sum_{j=1}^{n} (2j - 1)[\log t_j + \log(1 - t_{n-j+1})].$$

Durbin and Knott [50] suggested the statistic

$$(4.1.9) \qquad M_n^2 = \frac{n+1}{n} \sum_{j=1}^{n} \left(t_j - \frac{j}{n+1} \right)^2,$$

which bears a similar relation to W_n^2 as do the Pyke–Brunk statistics (2.3.1) and (2.3.2) to the Kolmogorov–Smirnov statistics (2.1.1)–(2.1.3).

4.2. Exact results. Little is known about the exact distributions of Cramér–von Mises statistics. Marshall [84] has given explicit expressions for the d.f.'s of W_1^2, W_2^2 and W_3^2 and Stephens and Maag [121] have given formulae for the extreme lower-tail probabilities for W_n^2. The exact first four moments of W_n^2 have been given by Pearson and Stephens [92]. In view of this small amount of knowledge available the following difference–differential equation for the characteristic function of a general class of statistics seems worth attention.

Apart from constants, all the statistics (4.1.5)–(4.1.9) take the form $\sum_{j=1}^{n} g_j(t_j)$. Let

$$(4.2.1) \quad \psi(\theta, j, t) = \int_0^t e^{ig_j(t_j)\theta} \int_0^{t_j} e^{ig_{j-1}(t_j - 1)\theta} \cdots \int_0^{t_2} e^{ig_1(t_1)\theta}\, dt_1 \cdots dt_j, \quad j = 1, \cdots n.$$

Then the characteristic function of $\sum_{j=1}^{n} g_j(t_j)$ is $n!\,\psi(\theta, n, 1)$. On differentiating (4.2.1) we obtain the difference–differential equation

$$(4.2.2) \qquad \frac{d\,\psi(\theta, j, t)}{dt} = \psi(\theta, j - 1, t)\, e^{i\theta g_j(t)}, \qquad j = 1, \cdots, n, \quad 0 < t \leq 1,$$

with boundary conditions $\psi(\theta, j, 0) = 0$, $j = 1, \cdots, n$, $\psi(\theta, 0, t) = 1$, $0 \leq t \leq 1$. This derivation is due to Martin Knott.

In general an analytical solution cannot be found and only numerical methods are available; (4.2.2) has to be solved for $n! \; \psi(\theta, n, 1)$ which then has to be inverted numerically. This would appear to be a formidable task, though not altogether beyond hope.

A relation similar to (4.2.2) can be obtained for the distribution function $p(y, j, t) = \Pr(y(t) \leq y | t_j < t < t_{j+1})$ on putting this equal to $j! q(y, j, t)/t^j$, that is,

$$(4.2.3) \qquad \frac{dq(y, j, t)}{dt} = q(y - g_j(t), j - 1, t), \qquad j = 1, \cdots, n, \quad 0 < t < 1,$$

but this seems less tractable than (4.2.2).

A method based on direct numerical evaluation of (4.2.1), the resulting values being used as coefficients in a Fourier series for the distribution function, has been used by Knott and me to obtain exact percentage points for W_n^2. A table of percentage points has also been given by Stephens and Maag [121] based in part on the exact results for the lower tail mentioned above, on a χ^2 approximation based on the exact first three moments and on Monte Carlo.

4.3. Components of Cramér-von Mises statistics. For each $\psi(t)$ the statistic

$$(4.3.1) \qquad \qquad \mathcal{W}_n^2 = \int_0^1 \psi(t) y_n(t)^2 \, dt,$$

where $y_n(t) = \sqrt{n}(F_n(t) - t)$, gives an overall measure of the discrepancy between observations and hypothesis. We shall now consider the problem of breaking down this general measure into a set of uncorrelated components, each measuring some distinctive aspect of the data.

Let $\xi_n(t) = \sqrt{(\psi(t))} y_n(t)$. Then $\mathcal{W}_n^2 = \int_0^1 \xi_n(t)^2 \, dt$. Also $E(\xi_n(t)) = 0$ and

$$(4.3.2) \qquad \rho(t, t') = E(\xi_n(t)\xi_n(t')) = \sqrt{(\psi(t)\psi(t'))} (\min(t, t') - tt').$$

We wish to represent $\mathcal{W}_n^2$ in the form $\sum_{j=1}^{\infty} \lambda_j \zeta_{nj}^2$, where the ζ_{nj} are uncorrelated random variables each with zero mean and unit variance. The way to do this can be understood by considering the analogous problem for the sum of squares $\sum_{j=1}^{N} \xi_j^2$ of a finite number of random variables $\xi_1, \cdots, \xi_N$ with zero means and positive-definite variance matrix V. It is well known that for this problem, $\sum_{j=1}^{N} \xi_j^2 = \sum_{j=1}^{N} \lambda_j \zeta_j^2$ with $\zeta_j = \lambda_j^{-1/2} l_j' \xi$, where λ_j, l_j, $j = 1, \cdots, N$, are the eigenvalues and orthonormalized eigenvectors of V, i.e., the orthonormalized solutions of

$$(4.3.3) \qquad \qquad V l = \lambda l,$$

where $\xi = [\xi_1, \cdots, \xi_N]'$ and where $\zeta_1, \cdots, \zeta_N$ are uncorrelated with zero means and unit variances. The inverse relation is $\xi_j = \sum_{i=1}^{N} l_{ij} \lambda_i^{1/2} \zeta_i$, where l_{ij} is the jth element of l_i.

Analogously, for the continuous case suppose that λ_j, $l_j(t)$, $j = 1, 2, \cdots$, are the eigenvalues and orthonormalized eigenfunctions of the kernel $\rho(t, t')$, i.e., the solutions of the integral equation

$$(4.3.4) \qquad \int_0^1 \rho(t, t') l(t') \, dt' = \lambda l(t)$$

satisfying $\int_0^1 l_j(t)^2 \, dt = 1$ and $\int_0^1 l_j(t) l_k(t) = 0, j \neq k$. Putting $\zeta_{nj} = \lambda_j^{-1/2} \int_0^1 l_j(t) \xi_n(t) \, dt$ we obtain

$$(4.3.5) \qquad W_n^2 = \sum_{j=1}^{\infty} \lambda_j \zeta_{nj}^2,$$

where $\zeta_{n1}, \zeta_{n2}, \cdots$ are uncorrelated random variables with zero means and unit variances. We also have the inverse relation

$$(4.3.6) \qquad \xi_n(t) = \sum_{j=1}^{\infty} \lambda_j^{1/2} \zeta_{nj} l_j(t).$$

We merely sketch the proof. Assuming that $\psi(t)$ is continuous, it follows that $\rho(t, t')$ is continuous and hence by Mercer's theorem can be expanded in the uniformly convergent series

$$\rho(t, t') = \sum_{j=1}^{\infty} \lambda_j l_j(t) l_j(t'),$$

where the l_j's are orthonormal. Now

$$E(\zeta_{nj} \zeta_{nk}) = \lambda_j^{-1/2} \lambda_k^{-1/2} \int_0^1 \int_0^1 l_j(t) l_k(t') \rho(t, t') \, dt \, dt' = \delta_j^k$$

and

$$E(\xi_n(t) \zeta_{nj}) = \lambda_j^{-1/2} \int_0^1 l_j(t') \rho(t, t') \, dt' = \lambda_j^{1/2} l_j(t).$$

Hence,

$$E \left| \xi_n(t) - \sum_{j=1}^{m} \lambda_j^{1/2} \zeta_{nj} l_j(t) \right|^2 = \rho(t, t) - \sum_{j=1}^{m} \lambda_j l_j(t)^2,$$

which $\rightarrow 0$ as $m \rightarrow \infty$ uniformly in t, so that $\sum_{j=1}^{\infty} \lambda_j^{1/2} \zeta_{nj} l_j(t)$ converges in mean square uniformly to $\xi_n(t)$. When ρ is positive definite the eigenfunctions form a complete orthonormal set and Parseval's theorem gives (4.3.5) immediately. Detailed treatments are given by Kac and Siegert [67], Rosenblatt [100, Chap. VIIIc] and Gikhman and Skorokhod [59, Chap. 5, § 2.3].

Now let us consider the solution of (4.3.4), i.e., of

$$(4.3.7) \qquad \int_0^1 \psi(t)^{1/2} \psi(t')^{1/2} (\min(t, t') - tt') l(t') \, dt' = \lambda l(t).$$

Put $h(t) = [\psi(t)]^{-1/2}l(t)$ when $\psi(t) \neq 0$ and $h(t) = 0$ when $\psi(t) = 0$. Substituting in (4.3.7) and splitting the range of integration at t we obtain

$$(1 - t)\int_0^t t'\psi(t')h(t')\,dt' + t\int_t^1 (1 - t')\psi(t')h(t')\,dt' = \lambda h(t).$$

Differentiating both sides twice and simplifying gives

(4.3.8) $$\lambda\frac{d^2h(t)}{dt^2} + \psi(t)h(t) = 0.$$

This has to be solved subject to the conditions $h(0) = h(1) = 0$ which we derive from (4.3.7).

As an example take the ordinary Cramér–von Mises statistic for which $\psi(t) = 1$ for all t. We find $\lambda_j = (j^2\pi^2)^{-1}$, $l_j(t) = \sqrt{2}\sin(j\pi t)$, $j = 1, 2, \cdots$. Putting

(4.3.9) $$z_{nj} = \sqrt{2}\,j\pi\int_0^1 \sin(j\pi t)y_n(t)\,dt,\qquad\qquad j = 1, 2, \cdots,$$

we obtain

(4.3.10) $$W_n^2 = \sum_{j=1}^\infty \frac{z_{nj}^2}{j^2\pi^2},$$

where z_{n1}, z_{n2}, $\cdots$ are uncorrelated with zero means and unit variances. The inverse of (4.3.9) is

(4.3.11) $$y_n(t) = \sqrt{2}\sum_{j=1}^\infty \frac{\sin(j\pi t)}{j\pi}z_{nj},\qquad\qquad 0 \leq t \leq 1.$$

Durbin and Knott [50] suggest that when using W_n^2 for testing goodness of fit the first few z_{nj}'s should be examined and tested also. They show that z_{nj} has the alternative form

(4.3.12) $$z_{nj} = \left(\frac{2}{n}\right)^{1/2}\sum_{i=1}^n \cos(j\pi t_i),$$

which can be written as a function of the unordered observations, i.e., if $t'_1, \cdots, t'_n$ are the original independent U(0, 1) observations, then $z_{nj} = \sqrt{(2/n)}\sum_{i=1}^n \cos(j\pi t')$. They show further that for each n the z_{nj}'s are identically distributed and give a table of significance points. They also consider the relation of the components to a Fourier analysis of the data and examine the components of the Anderson–Darling statistic A_n^2. These turn out to be essentially the same as the Legendre polynomial functions used by Neyman [89] as the basis for the construction of his "smooth" test of goodness of fit. For further details the reader is referred to the Durbin–Knott paper.

4.4. Asymptotic distributions. In this section we consider the asymptotic distribution of the generalized Cramér–von Mises statistic $\mathscr{W}_n^2 = \int_0^1 \psi(t)y_n(t)^2\,dt$. Since $\{y_n(t)\} \xrightarrow{\mathscr{D}} \{y(t)\}$, where $\{y(t)\}$ is the tied-down Brownian motion process, it follows from the treatment of § 3.2 that $W_n^2 \xrightarrow{\mathscr{D}} W^2 = \int_0^1 \psi(t)y(t)^2 \, dt$ whenever W_n^2 is continuous in metric d.

For consideration of the continuity of W_n^2 we confine ourselves for simplicity to the case where $\psi(t)$ is continuous for $0 \leq t \leq 1$. It is then easy to show that $z_n(t) = \psi(t)y_n(t)^2$, when regarded as a function of $y_n(t)$, is continuous in d so $z_n(t) \xrightarrow{\mathscr{D}} z(t) = \psi(t)y(t)^2$. The result $W_n^2 \xrightarrow{\mathscr{D}} W^2$ will then follow if we can show that $\int_0^1 z(t) \, dt$ as a function of $z(t)$ is continuous in d. Now for any sequence of functions z_m converging to z in d there exist functions λ_m such that $\lim_m z_m(\lambda_m(t)) = z(t)$ uniformly in t and $\lim_m \lambda_m(t) = t$ uniformly in t, cf. Billingsley [11, p. 112]. Since every element of D is bounded and has at most a countable number of discontinuities it is Riemann integrable. Take Riemann subdivisions $0 < t_1 < \cdots < t_p < 1$ for $\int_0^1 z(t) \, dt$ and $0 < \lambda_m(t_1) < \cdots < \lambda_m(t_p) < 1$ for $\int_0^1 z_m(t) \, dt$. As $m \to \infty$ the upper and lower sums for the latter integral converge to those for the former. It follows that $\int_0^1 z_m(t) \, dt \to \int_0^1 z(t) \, dt$ and hence that $\int_0^1 z(t) \, dt$ is continuous in d.

It is important to note that this argument does not apply to some cases important in practice, notably the Anderson–Darling statistic $A_n^2 = \int_0^1 [t(1-t)]^{-1} \cdot y_n(t)^2 \, dt$, since the function $[t(1-t)]^{-1}$ is not continuous at $t = 0$ or 1. For this case one could either consider first the convergence of the statistic obtained by integrating over the range $(\delta, 1-\delta)$ and then let $\delta \to 0$ or alternatively, and simpler, use a direct argument based on the orthogonal form of A_n^2 employing Rosenblatt's [99] method of proof.

Now we consider the asymptotic distribution of the basic Cramér–von Mises statistic W_n^2. From the above this is the same as the distribution of

$$(4.4.1) \qquad\qquad W^2 = \int_0^1 y(t)^2 \, dt,$$

where $\{y(t)\}$ is the tied-down Brownian motion process. Since the eigenvalues of the kernel $\min(t, t') - tt'$ are $(j^2\pi^2)^{-1}$ and the eigenfunctions are $\sqrt{2}\sin(j\pi t)$, $j = 1, 2, \cdots$, the treatment of the previous section implies

$$(4.4.2) \qquad\qquad W^2 = \sum_{j=1}^{\infty} \frac{z_j^2}{j^2\pi^2},$$

where

$$(4.4.3) \qquad z_j = \sqrt{2}\, j\pi \int_0^1 \sin(j\pi t)y(t) \, dt, \qquad\qquad j = 1, 2, \cdots,$$

(cf. (4.3.9) and (4.3.10)). It is easily shown directly from (4.4.3) that $z_1, z_2, \cdots$ are independent $N(0, 1)$ variables. A rigorous proof that the distribution of (4.4.1) is the same as that of (4.4.2) is given by Hájek and Sidák [63, § V 3.3].

Now consider the random variable $W_{(m)}^2 = \sum_{j=1}^{m} z_j^2/(j^2\pi^2)$. This converges in distribution to W^2 as $m \to \infty$ and its characteristic function is

$$\prod_{j=1}^{m} \left(1 - \frac{2i\theta}{j^2\pi^2}\right)^{-1/2}.$$

The characteristic function of W^2 is therefore

$$\phi(\theta) = \prod_{j=1}^{\infty} \left(1 - \frac{2i\theta}{j^2\pi^2}\right)^{-1/2}$$

(4.4.4)

$$= \left(\frac{\sqrt{(2i\theta)}}{\sin\sqrt{(2i\theta)}}\right)^{1/2}$$

in virtue of the infinite-product form of the sine function. This was inverted by Smirnov [105], [106] to give

(4.4.5) $\displaystyle \Pr(W^2 \le x) = 1 - \frac{1}{\pi} \sum_{j=1}^{\infty} (-1)^{j-1} \int_{(2j-1)^2\pi^2}^{(2j)^2\pi^2} y^{-1} \left(\frac{-\sqrt{y}}{\sin\sqrt{y}}\right)^{1/2} e^{-xy/2} \, dy.$

A simpler derivation of this inversion, together with suggestions for techniques of numerical evaluation is given by Slepian [104] for a problem which, although arising in a different context from the above, is mathematically the same in essentials.

A different inversion of (4.4.4) was given as a rapidly-converging series by Anderson and Darling [4] and was used by them to tabulate the distribution function of W^2.

Similarly, the Anderson–Darling statistic A_n^2 converges in distribution to

(4.4.6) $\displaystyle A^2 = \int_0^1 \frac{y(t)^2 \, dt}{t(1 - t)},$

where $y(t)$ is tied-down Brownian motion, and this has characteristic function

$$\phi_A(\theta) = \prod_{j=1}^{\infty} \left(1 - \frac{2i\theta}{j(j + 1)}\right)^{-1/2}$$

since the eigenvalues of the kernel $[t(1 - t)t'(1 - t')]^{-1/2}(\min(t, t') - tt')$ are $[j(j + 1)]^{-1}$, $j = 1, 2, \cdots$, (Anderson and Darling [4]). Using the technique described in Whittaker and Watson [135, § 7.5], this can be put in the alternative form

(4.4.7) $\displaystyle \phi_A(\theta) = \left[\frac{-2\pi i\theta}{\cos(\pi\sqrt{(1 + 8i\theta)/2})}\right]^{1/2}.$

Anderson and Darling gave a rapidly converging series for the inverse of $\phi_A(\theta)$ and used this in [5] to give a short table of significance points.

Darling [35] has given the first two terms of an expansion of the Laplace transform of W_n^2 in powers of n^{-1}.

4.5. Other statistics based on the sample d.f. Statistics of the Kolmogorov–Smirnov and Cramér–von Mises types statistics are not the only test statistics based on the sample d.f. Indeed, the chi-squared test itself is in this category since for group boundaries $X_1, \cdots, X_{k-1}$ the chi-squared statistic is

$$X^2 = n \sum_{i=1}^{k} \frac{[F_n(X_i) - F_n(X_{i-1}) - F(X_i) + F(X_{i-1})]^2}{F(X_i) - F(X_{i-1})},$$

where $X_0 = -\infty$ and $X_k = +\infty$. Moreover, with suitable interpretation the wide class of tests based on the sample spacings could be regarded as falling within the category, or at least as very closely related. We shall not consider these further tests. For tests based on spacings we refer the reader to the review article by Pyke [94]. Here we shall include because of its simplicity only one further statistic, the mean of the sample d.f., that is,

$$(4.5.1) \qquad \bar{F}_n = \int_0^1 F_n(t)\, dt = 1 - \bar{t} \quad \text{where} \quad \bar{t} = \frac{1}{n} \sum_{i=1}^n t_i.$$

Thus $1 - \bar{F}_n$ is the mean of n independent $U(0, 1)$ variables and so converges rapidly to normality with mean $\frac{1}{2}$ and variance $1/(12n)$. This statistic was suggested by Birnbaum and Tang [16]. Exact significance points have been tabulated by Stephens [116].

5. Tests on the circle.

5.1. Kuiper's statistic V_n. In a number of sciences, notably biology and geophysics, situations occur where the basic data are in the form of directions and one wishes to test the hypothesis that the directions are oriented at random. For example, the data might consist of the directions in which a sample of birds fly off when released at a fixed point. If one imagines a circle of unit circumference centered at this point, then the data can be represented as a set of n points on the circle. The test of the hypothesis of randomness of directions is then equivalent to testing the hypothesis that the n points are distributed at random on the circumference of the circle.

Let P be an arbitrary point on the circumference and let $0 \leq t_1 \leq \cdots \leq t_n \leq 1$ be the distances from P of the n points as one travels round the circle in a fixed sense, say anti-clockwise. Since when P is fixed $t_1, \cdots, t_n$ determine the sample completely, the test statistic should depend on these values only. But since P is arbitrary it is desirable that the statistic should remain invariant under choice of P.

Many such statistics could be devised but we shall confine ourselves to statistics derived from the sample d.f. Of these the first was suggested by Kuiper [77] and is the statistic of Kolmogorov–Smirnov type appropriate for the circle. For arbitrary P let $t_1, \cdots, t_n$ be as above and let $F_n(t)$ be the sample d.f. computed from these values. Kuiper's statistic is

$$(5.1.1) \qquad V_n = D_n^+ + D_n^-,$$

where D_n^+ and D_n^- are as before.

Our first task is to show that the value of V_n does not depend on the choice of P. Let $\tilde{P}$ be a new origin and let $\tilde{D}_n^+$, $\tilde{D}_n^-$ be the new values of D_n^+, D_n^-. Putting $\tilde{l}$ equal to the length of the arc $P\tilde{P}$ when measured in an anti-clockwise direction, we shall consider the changes in $\tilde{D}_n^+$ and $\tilde{D}_n^-$ as $\tilde{l}$ moves from 0 to 1 keeping P fixed.

It is easy to check that for $t_i \leqq \tilde{\imath} < t_{i+1}$ we have $\tilde{D}_n^+ = D_n^+ - i/n + \tilde{\imath}$ and $\tilde{D}_n^-$ $= D_n^- + i/n - \tilde{\imath}$, $i = 0, \cdots, n$, $t_0 = 0$, $t_{n+1} = 1$. Thus $\tilde{D}_n^+ + \tilde{D}_n^- = D_n^+ + D_n^-$ for all $\tilde{\imath}$, i.e., V_n does not depend on the choice of P. It will be noted that this result does not depend on any distributional assumptions.

Before considering the distribution of V_n it should be emphasized that the practical use of this statistic is not restricted to tests on the circle. It can also be used for tests on the line and is indeed the appropriate statistic to use rather than D_n when one is more interested in discrimination against shifts in scale than against shifts in location.

5.2. Exact distribution of V_n. Our technique for finding the exact distribution of V_n will be to put the problem into a form which enables us to exploit results already given in § 2.4. Let $A_1, \cdots, A_n$ be the n points on the circle determined by a particular sample. We may choose the origin P arbitrarily but as it is moved round the circle the particular point A_j at which $D_n^- = \max(t_j - (j - 1)/n)$ is attained remains fixed; call this A^* and assume it to be unique which it will be with probability one.

Consider a random mechanism for the choice of P under which, for every sample of points, P is located at $A_1, \cdots, A_n$ with equal probabilities. Our starting point is the assertion that $\Pr(V_n \leqq v) = \Pr(V_n \leqq v | P = A^*)$. This follows since the value of V_n is unaffected by the choice of P.

Now

$$
\Pr(V_n \leqq v | P = A^*) = \frac{\Pr(V_n \leqq v \text{ and } P = A^* | P = A_1 \text{ or } \cdots \text{ or } A_n)}{\Pr(P = A^* | P = A_1 \text{ or } \cdots \text{ or } A_n)}
$$

(5.2.1)

$$
= n \Pr(V_n \leqq v \text{ and } P = A^* | P = A_1 \text{ or } \cdots \text{ or } A_n).
$$

Consider the graph of $F_n(t)$. If $P = A_1$ or $\cdots$ or A_n, there is always a jump of $1/n$ at $t = 0$. Once P is fixed the remaining $n - 1$ points constitute a random sample of $n - 1$ points on the interval $(0, 1)$. These two facts together imply that the progress of the graph from the point $(0, 1/n)$ to the point $(1, 1)$ corresponds to that of the graph of $F_{n-1}(t)$ for a random sample of $n - 1$ points on $(0, 1)$ with the exception that jumps of $1/n$ are taken instead of $1/(n - 1)$.

If, further, $P = A^*$, then the graph of $F_n(t)$ must not fall below the line $y = t$. Finally, if $V_n \leqq v$, then the graph of $F_n(t)$ must not cross the line $y = v + t$. Thus the probability that $V_n \leqq v$ and $P = A^*$ given that $P = A_1$ or $\cdots$ or A_n is the same as the probability that the graph of $F_n(t)$ lies between the lines $y = t$ and $y = t + v$ given that there is a jump of $1/n$ at the origin. Switching to the sample of $n - 1$ points and scaling up the jumps from $1/n$ to $1/(n - 1)$, this is the same as the probability that the graph of $F_{n-1}(t)$ lies between the lines

$$
(5.2.2) \qquad y = -\frac{1}{n-1} + \frac{n}{n-1}t \quad \text{and} \quad y = \frac{nv - 1}{n-1} + \frac{n}{n-1}t.
$$

But this is simply the quantity $p_{n-1}(nv - 1, 1, 1)$, several methods for the computing of which have been given in § 2.4. We therefore have finally

(5.2.3) $$\Pr(V_n \leq v) = np_{n-1}(nv - 1, 1, 1),$$

where $p_n(a, b, c)$ is the probability that the graph of $F_n(t)$, based on a random sample of n points on $(0, 1)$, lies between the lines $ny = a + (n + c)t$ and $ny = -b + (n + c)t$ with $a, b, a + c, b - c > 0$.

Putting

(5.2.4) $$q_r = \frac{(r + 1)^r}{r!} p_r(nv - 1, 1, 1)$$

we can immediately write down from (2.4.12) the difference equation for q_r:

(5.2.5) $$\sum_{j=0}^{[nv]} (-1)^j \frac{(nv - j)^j}{j!} q_{r-j} = 0, \qquad r = [nv], [nv] + 1, \cdots.$$

Since $p_r(nv - 1, 1, 1) = (r + 1)^{-1}$ for $r < [nv]$ by (2.1.7), the initial conditions are

(5.2.6) $$q_0 = 1, \quad q_r = \frac{(r + 1)^{r-1}}{r!}, \qquad r = 1, \cdots, [nv] - 1.$$

Solving (5.2.5) and using (5.2.3) and (5.2.4) we obtain

(5.2.7) $$\Pr(V_n \leq v) = \frac{(n - 1)!}{n^{n-2}} q_{n-1}, \qquad n = 2, 3, \cdots.$$

The exact distribution of V_n was first considered by Stephens [114] who obtained (5.2.5) for $[nv] = 1$ and $[nv] = 2$ and gave further results for the upper-tail probabilities. In fact, (5.2.5) holds for all values of $v > 1/n$.

For practical calculations this difference-equation method is liable to rounding error and for larger values of n it may be preferable to obtain the value of $p_{n-1}(nv - 1, 1, 1)$ from the appropriate element (in this case the top left-hand element) of the nth power of the matrix H_1 in the way described in § 2.4. For the present application the matrix H_1 in (2.4.10) has $p = [nv] + 1$, $\delta = 0$, $\varepsilon = [nv] + 1 - nv$ and $h = 0$.

Brunk [72] suggested the statistic

(5.2.8) $$B_n = C_n^+ + C_n^-,$$

where C_n^+ and C_n^- are the Pyke statistics defined in (2.3.1) and (2.3.2), as an alternative to V_n and obtained recursions for the calculation of its exact distribution. Stephens [118] proved that $\Pr(B_n \leq b) = \Pr(V_n \leq b + 1/(n + 1))$. Thus the above methods may be used to compute the distribution function of B_n.

5.3. Asymptotic distribution of V_n. From (3.4.10) we know that

(5.3.1) $$\lim_{n \to \infty} \Pr(\sqrt{n} D_n^+ \leq x, \sqrt{n} D_n^- \leq y) = \sum_{j=-\infty}^{\infty} \{e^{-2j^2(x+y)^2} - e^{-2[jx+(j-1)y]^2}\}.$$

Denoting this by $\phi(x, y)$ we have

$$(5.3.2) \quad \lim_{n \to \infty} \Pr(\sqrt{n} \, V_n \leq v) = \int_0^v \int_0^{v-y} \frac{\partial^2 \phi(x, y)}{\partial x \, \partial y} \, dx \, dy = \int_0^v \left(\frac{\partial \phi(x, y)}{\partial y} \bigg|_{x = v - y} \right) dy.$$

Now

$$\frac{\partial \phi}{\partial y} \bigg|_{x = v - y} = \sum_{j = -\infty}^{\infty} \{ -4j^2 v \, e^{-2j^2 v^2} + 4(j - 1)(jv - y) \, e^{-2(jv - y)^2} \},$$

which on integrating with respect to y from 0 to v gives

$$(5.3.3) \quad \lim_{n \to \infty} \Pr(\sqrt{n} \, V_n \leq v) = 1 - 2 \sum_{j=1}^{\infty} (4j^2 v^2 - 1) \, e^{-2j^2 v^2},$$

the result given by Kuiper [78]. Kuiper also obtained the second term in the series expansion of $\Pr(\sqrt{n} \, V_n \leq v)$ giving

$$\Pr(\sqrt{n} \, V_n \leq v) = 1 - 2 \sum_{j=1}^{\infty} (4j^2 v^2 - 1) \, e^{-2j^2 v^2}$$

$$(5.3.4)$$

$$+ \frac{8}{3\sqrt{n}} \sum_{j=1}^{\infty} j^2 (4j^2 v^2 - 3) \, e^{-2j^2 v^2} + O\left(\frac{1}{n} \right).$$

5.4. Watson's statistic U_n^2. We now consider the statistic of Cramér–von Mises type appropriate for tests on the circle. For testing a hypothetical d.f. $F_0(x)$ Watson [129] suggested the statistic

$$U_n^2 = n \int_{-\infty}^{\infty} [F_n(x) - F_0(x) - \overline{F_n(x) - F_0(x)}]^2 \, dF_0(x)$$

$$(5.4.1) \qquad = n \int_0^1 [F_n(t) - t - \overline{F_n(t) - t}]^2 \, dt$$

$$= \int_0^1 (y_n(t) - \bar{y}_n)^2 \, dt = \int_0^1 y_n(t)^2 \, dt - \bar{y}_n^2$$

on putting $t = F_0(x)$, $y_n(t) = \sqrt{n}(F_n(t) - t)$, $\overline{F_n(x) - F_0(x)} = \int_{-\infty}^{\infty} (F_n(x) - F_0(x)) \, dF_0(x)$ and similarly for $\overline{F_n(t) - t}$ and $\bar{y}_n$. Hence we obtain the form more suitable for practical work:

$$(5.4.2) \quad U_n^2 = W_n^2 - n(\bar{t} - \tfrac{1}{2})^2 = \sum_{j=1}^{n} \left(t_j - \frac{j - \frac{1}{2}}{n} \right)^2 - n(\bar{t} - \tfrac{1}{2})^2 + \frac{1}{12n},$$

where $\bar{t} = n^{-1} \sum_{j=1}^{n} t_j$. As before, an origin P is chosen at an arbitrary point on the circle prior to the calculation of $F_n(x)$ or $F_n(t)$. Although Watson introduced this statistic specifically for tests on the circle it can, like V_n, be used for tests on the line and indeed can be expected to be more powerful than W_n^2 for tests against shifts in variance of a symmetric distribution.

In order to show that the value of U_n^2 is independent of P, let P' be a new origin located at $t = t_0$ relative to origin P. Let $t' = t - t_0$, $t_0 \leq t \leq 1$, and $t' = 1 + t - t_0$, $0 \leq t < t_0$, be the new time-variable and let $y_n'(t')$ be the new sample process. Then $y_n'(t') = y_n(t_0 + t') - y_n(t_0)$, $t_0 \leq t \leq 1$, and $y_n'(t') = y_n(t_0 + t' - 1) - v_n(t_0)$, $0 \leq t \leq t_0$, while the new mean is $\bar{y}_n' = \bar{y}_n - y_n(t_0)$. On taking differences the term $y_n(t_0)$ disappears, so the new differences $y_n'(t') - \bar{y}_n$ are merely a cyclic transformation of the old. Thus $\int_0^1 (y_n(t) - \bar{y}_n)^2 \, dt = \int_0^1 (y_n'(t) - \bar{y}_n')^2 \, dt$, i.e., the value of U_n^2 is unaffected by the choice of origin.

5.5. Exact results for U_n^2. A detailed study of exact results for U_n^2 has been made by Stephens [112], [113] who gives the exact first four moments, the exact distribution of U_n^2 for $n = 1, 2, 3, 4$, some exact results for the lower tail of the distribution for any n and tables of significance points for both upper and lower tails. The algebra required for these exact results is too lengthy to be included here. For those parts of the range where exact results were not available, Pearson curves based on the first four moments were used by Stephens to obtain approximate significance points.

5.6. Asymptotic distribution of U_n^2. Let

$$(5.6.1) \qquad U^2 = \int_0^1 (y(t) - \bar{y})^2 \, dt = \int_0^1 y(t)^2 \, dt - \bar{y}^2,$$

where $\{y(t)\}$ is the tied-down Brownian motion process and $\bar{y} = \int_0^1 y(t) \, dt$. It was shown in § 4.4 that W_n^2 and $\bar{y}_n$ when regarded as functions of $y_n(t)$ are continuous in metric d. Thus $U_n^2 = W_n^2 - \bar{y}_n^2$ is continuous in d from which the result $U_n^2 \xrightarrow{\mathscr{D}} U^2$ follows immediately.

Let $u(t) = y(t) - \bar{y}$. The covariance kernel of $u(t)$ is

$$(5.6.2) \qquad \rho(t, t') = E(u(t)u(t')) = \min(t, t') - \tfrac{1}{2}(t + t') + \tfrac{1}{2}(t - t')^2 + \tfrac{1}{12},$$

$$0 \leq t, t' \leq 1.$$

Putting $t = t'$ we see that the variance is equal to the constant value of $\frac{1}{12}$ for all t. This means that there is no point in attempting any weighting of the Anderson–Darling [4] type.

In order to express U^2 as a weighted sum of squares similar to the expression (4.4.2) for W^2 we need the eigenvalues and orthonormalized eigenfunctions of (5.6.2), i.e., the orthonormal solutions of

$$(5.6.3) \qquad \int_0^1 \rho(t, t')h(t') \, dt' = \lambda h(t), \qquad 0 \leq t \leq 1.$$

Inspection shows that $h(t) \equiv 1$, $\lambda = 0$ is a solution. Let $h(t)$ be any other eigenfunction. Since this must be orthogonal to $h(t) \equiv 1$ the condition

$$(5.6.4) \qquad \int_0^1 h(t) \, dt = 0$$

must be satisfied. We observe further that $\lambda h(0) = \int_0^1 (-\frac{1}{2}t' + \frac{1}{2}t'^2 + \frac{1}{12})h(t')\,dt'$ $= \lambda h(1)$ which gives the further condition

(5.6.5) $$h(0) = h(1), \qquad \lambda \neq 0.$$

Splitting the range of integration at t, differentiating twice and using (5.6.4) we obtain

(5.6.6) $$\lambda \frac{d^2 h(t)}{dt^2} + h(t) = 0$$

whose solutions satisfying (5.6.4) and (5.6.5) are

(5.6.7)
$$\lambda_{2j-1} = \lambda_{2j} = \frac{1}{4j^2\pi^2},$$
$$h_{2j-1}(t) = \sqrt{2}\sin(2\pi jt), \quad h_{2j}(t) = \sqrt{2}\cos(2\pi jt),$$
$$j = 1, 2, \cdots.$$

Analogously to (4.4.2) we therefore obtain

(5.6.8) $$U^2 = \sum_{j=1}^{\infty} \frac{z_{2j-1}^2 + z_{2j}^2}{4j^2\pi^2},$$

where $z_1, z_2, \cdots$ are independent $N(0, 1)$ variables. Comparing (5.6.8) and (4.4.2) we observe the interesting fact that $4U^2$ has the same distribution as the sum of two independent values of W^2.

Putting $\frac{1}{2}(z_{2j-1}^2 + z_{2j}^2) = w_j$ we obtain

$$U^2 = \sum_{j=1}^{\infty} \frac{w_j}{2j^2\pi^2},$$

where $w_1, w_2, \cdots$ are independent exponentially distributed variables and this has characteristic function

(5.6.9)
$$\phi(\theta) = E(e^{i\theta U^2}) = \prod_{j=1}^{\infty} \left(1 - \frac{i\theta}{2j^2\pi^2}\right)^{-1}$$
$$= \frac{\sqrt{(\frac{1}{2}i\theta)}}{\sin\sqrt{(\frac{1}{2}i\theta)}}$$

by the infinite product for the sine function. Expressing (5.6.9) in partial fractions gives

$$\phi(\theta) = 2 \sum_{j=1}^{\infty} \frac{(-1)^{j-1}}{1 - i\theta/(2j^2\pi^2)}$$

from which after inversion we obtain finally

(5.6.10) $$\Pr(U^2 \leq u) = \sum_{j=-\infty}^{\infty} (-1)^{j-1} e^{-2j^2\pi^2 u} = \lim_{n \to \infty} \Pr(U_n^2 \leq u).$$

This derivation is due to Watson [129].

Comparing (5.6.10) with expression (3.4.8) for the limiting d.f. of Kolmogorov's statistic we observe the remarkable fact that with the substitution $d^2 = \pi^2 u$ the two expressions are the same. This means that nD_n^2 and $\pi^2 U_n^2$ have the same limiting distribution. What is even more bizarre, it means that $4nD_n^2/\pi^2$ has the same limiting distribution as the sum of two independent values of W_n^2. No simple explanation has so far been given for this interesting phenomenon.

5.7. Other statistics for tests on the circle. Let $N(t)$ be the number of points within a semi-circular arc between the points $t, t + \frac{1}{2}$ on a circle of unit circumference. Ajne [2] suggested the statistic

(5.7.1)
$$N_n = \max_t N(t)$$

for testing the hypothesis that n points are distributed randomly on the circle. Putting $N_n^* = (2N_n - n)/\sqrt{n}$ he showed that when the hypothesis is true,

(5.7.2)
$$\lim_{n \to \infty} \Pr(N_n^* \geq c) = \lim_{n \to \infty} \Pr\left(\sqrt{n}\,D_n \leq \frac{\pi}{2c}\right),$$

where D_n is the ordinary Kolmogorov statistic. Since for testing significance one requires the upper tail of the distribution of N_n^*, it is the lower-tail form (3.4.9) of Kolmogorov's limiting distribution that is relevant. Thus we obtain

(5.7.3)
$$\lim_{n \to \infty} \Pr(N_n^* \geq c) = 2c\left(\frac{2}{\pi}\right)^{1/2} \sum_{j=0}^{\infty} e^{-(2j+1)^2 c^2/2}.$$

Following up a suggestion of Ajne, Watson [131] considered the related Cramér–von Mises type statistic

(5.7.4)
$$a_n^2 = n^{-1} \int_0^1 |N(t) - \tfrac{1}{2}n|^2\, dt$$

and by methods similar to those used for the limiting distribution of U_n^2 showed that

(5.7.5)
$$\lim_{n \to \infty} \Pr(a_n^2 > a) = \frac{2}{\pi} \sum_{j=0}^{\infty} \frac{(-1)^j}{(j + \frac{1}{2})} e^{-2(j + 1/2)^2 \pi^2 a}.$$

Further results on these and related statistics are given by Watson [132], Stephens [117] and Beran [9], [10].

6. Two-sample tests.

6.1. Kolmogorov–Smirnov tests. In this chapter we consider the situation where we have two independent samples of n, m independent observations from distributions with continuous distribution functions $F(x)$, $G(x)$ and we wish to test the hypothesis $H_0: F = G$, the common distribution being otherwise unspecified. There is an extremely large literature on tests of this hypothesis and Hájek and

Sidák [63] present a comprehensive treatment at an advanced technical level. We shall not attempt a general review but will confine ourselves to the study of the natural two-sample analogues of statistics already considered for the one-sample case. A wide variety of ad hoc combinatorial techniques have been employed for the solution of the resulting distribution problems. In contrast our approach will be to set the problems up in a form in which there is a natural analogy to the corresponding one-sample problems. We shall find that many of the techniques we employed in the one-sample case have very close counterparts in the two-sample situation.

We begin by considering the finite-sample distributions of the Kolmogorov–Smirnov statistics suggested by Smirnov [107], that is,

$$D_{nm}^+ = \sup_{-\infty \le x \le \infty} (F_n(x) - G_m(x)),$$

(6.1.1)
$$D_{nm}^- = \sup_{-\infty \le x \le \infty} (G_m(x) - F_n(x)),$$

$$D_{nm} = \max (D_{nm}^+, D_{nm}^-),$$

where $F_n(x)$, $G_m(x)$ are the sample d.f.'s of the two samples. Suppose that members of the first sample are denoted by A and members of the second sample by B. As x increases from $-\infty$ to $+\infty$, each time an A is met $F_n(x) - G_m(x)$ increases by $1/n$ and each time a B is met $F_n(x) - G_m(x)$ decreases by $1/m$. Let n_i be the number of A's which are $\le$ the ith observation in the combined sample. Then at the ith value,

(6.1.2)
$$F_n(x) - G_m(x) = \frac{n_i}{n} - \frac{i - n_i}{m} = \frac{n + m}{m} \left(\frac{n_i}{n} - \frac{i}{n + m} \right).$$

This form suggests the following two-sample analogue of the one-sample d.f. Let the observations in the combined sample be denoted by $x_1 \le \cdots \le x_{n+m}$ and let

(6.1.3)
$$F_{nm}(i) = \frac{n_i}{n},$$

where n_i is the number of A's $\le x_i$. Then $F_n(x) - G_m(x)$ takes values $[(n + m)/n]$ $\cdot [F_{nm}(i) - i/(n + m)]$, $i = 1, \cdots, n + m$, so

$$D_{nm}^+ = \frac{n + m}{m} \max_{i = 1, \cdots, n+m} \left(F_{nm}(i) - \frac{i}{n + m} \right),$$

(6.1.4)
$$D_{nm}^- = \frac{n + m}{m} \max_{i = 1, \cdots, n+m} \left(\frac{i}{n + m} - F_{nm}(i) \right),$$

$$D_{nm} = \max (D_{nm}^+, D_{nm}^-).$$

Obviously these statistics are distribution-free; indeed they depend only on the ranks of the members of one sample in the combined sample. Letting $m \to \infty$ and keeping n fixed, $D_{nm}^+ \to \sup (F_n(t) - t) = D_n^+$. We see that transforming to the ranks in the two-sample case is entirely analogous to transforming to the uniform distribution in the one-sample case. Now for D_n^+ we have the alternative form $D_n^+ = \max_j (j/n - t_j)$. The analogue of t_j is $r_j/(m + n)$, where r_j is the rank in the combined sample of the jth observation of the A sample. This gives the alternative form

$$(6.1.5) \qquad D_{nm}^+ = \frac{n + m}{m} \max_{j = 1, \cdots, n} \left(\frac{j}{n} - \frac{r_j}{n + m} \right).$$

Similarly we find

$$(6.1.6) \qquad D_{nm}^- = \frac{n + m}{m} \max_{j = 1, \cdots, n} \left(\frac{r_j - 1}{n + m} - \frac{j - 1}{n} \right)$$

as the analogue of $D_n^- = \max_j (t_j - (j - 1)/n)$. We get $r_j - 1$ not r_j here since the maximum is attained just *before* an A is reached, not *at* an A. Alternatively, if $s_1, \cdots, s_m$ are the ranks of the members of the second sample in the combined sample we deduce by symmetry from (6.1.1) and (6.1.5),

$$(6.1.7) \qquad D_{nm}^- = \frac{n + m}{n} \max_{j = 1, \cdots, m} \left(\frac{j}{m} - \frac{s_j}{n + m} \right).$$

Small-sample distributions of these statistics are best handled by Markov process techniques analogous to those used in the one-sample case. Just as there we took $F_n(t)$ as equivalent to $P_n(t)$ given $P_n(1) = 1$, here we take $F_{nm}(i)$ as equivalent to a path of a binomial process $\{B_n(i)\}$ with probability of success constant (for convenience we take this probability to be one half) and with jumps of $1/n$, conditional on $B_n(n + m) = 1$. Thus $nB_n(i)$ is the number of successes in the first i trials. We shall use this representation to obtain the distribution of a general class of statistics including all three statistics (6.1.1).

Consider the graph of $F_{nm}(i)$ as a function of i. If $D_{nm}^+ \leqq d$, then all points $(i, F_{nm}(i))$ must lie below or on the line $y = (md + i)/(n + m)$. Similarly if $D_{nm}^- \leqq d$, all points $(i, F_{nm}(i))$ must lie above or on the line $y = (-md + i)/(n + m)$. Let $a_j = \max \{i : j/n > (md + i)/(n + m)\}, j > n(md + 1)/(n + m)$ and let $b_j = \min \{i : j/n < (-md + i)/(n + m)\}$, $j < n - nmd/(n + m)$. Then the graph of $F_{nm}(i)$ will cross the upper line or the lower line or both if and only if it passes through at least one of the points $(a_j, j/n)$, $(b_j, j/n)$. Let these points be labelled $A_1, A_2, \cdots, A_q$ in order of increasing i, with the convention that if there are two such points for the same i they are labelled in increasing order of j. Then $D_{nm} > d$ if and only if the graph of $F_{nm}(i)$ passes through one or more of $A_1, \cdots, A_q$. We shall obtain recursive formulae which enable the probability of this to be computed. Our method is quite general and permits us to compute the probability of passing through one or more of any arbitrary set of points of the form $(i, j/n), i \in 1, \cdots, n + m$ and $j \in 0, \cdots, n$, provided that the points are labelled $A_1, \cdots, A_q$ in order of increasing i, and in order of increasing j where there is more than one point for the same value of i. This technique would therefore enable us to compute the distributions of weighted statistics analogous to those suggested for the one-sample case by Anderson and Darling [4].

As i increases from 1 to $n + m$, let p_r be the probability that a sample path of $\{B_n(i)\}$ passes through $A_r, r = 1, \cdots, q$, and let α_r be the probability that it passes through A_r, not having passed through any of $A_1, \cdots, A_{r-1}, r = 2, \cdots, q$ with $\alpha_1 = p_1$. Let p_{rs} be the probability that it passes through A_s given that it passes through $A_r, s > r$. Then, because of the strong Markov property of $\{B_n(i)\}$, we have

$$(6.1.8) \qquad p_s = \alpha_s + \sum_{r=1}^{s-1} \alpha_r p_{rs}, \qquad s = 2, \cdots, q, \quad p_1 = \alpha_1.$$

Denoting the point $(n + m, 1)$ by A_{q+1}, the probability that $B_n(i)$ passes through at least one of $A_1, \cdots, A_q$ and then passes through A_{q+1} is $\sum_{r=1}^{q} \alpha_r p_{r,q+1}$. Now the unconditional probability that $B_n(i)$ passes through A_{q+1} is

$$\binom{n+m}{m} 2^{-n}.$$

Thus the probability that $F_{nm}(i)$ passes through at least one of $A_1, \cdots, A_q$ is

$$(6.1.9) \qquad P_A = \frac{n!\,m!}{(n+m)!} 2^n \sum_{r=1}^{q} \alpha_r p_{r,q+1},$$

where this is evaluated as the conditional probability that $B_n(i)$ passes through at least one of $A_1, \cdots, A_q$ given that it passes through A_{q+1}.

We see that from a theoretical point of view the solution to the problem of finding the probability that $F_{nm}(i)$ crosses an arbitrary boundary is extremely simple. All that has to be done is solve the recursions (6.1.8) for $\alpha_1, \cdots, \alpha_q$ and substitute in (6.1.9). To complete the solution we need the values of p_r, p_{rs}. Suppose that A_s has co-ordinates $(i_s, j_s/n)$. Then p_s is just the probability of j_s successes in i_s trials, that is, $\binom{i_s}{j_s} 2^{-i_s}$ and p_{rs} is the probability of $j_s - j_r$ successes in $i_s - i_r$ trials, that is, it is

$$\binom{i_s - i_r}{j_s - j_r} 2^{-(i_s - i_r)} \quad \text{for} \quad i_s - i_r \geq j_s - j_r \quad \text{and zero otherwise};$$

also

$$p_{r,q+1} = \binom{n + m - i_r}{n - j_r} 2^{-(n+m-i_r)}.$$

Putting $a_r = 2^{i_r} \alpha_r$, we obtain the recursions

$$(6.1.10) \qquad a_s = \binom{i_s}{j_s} - \sum_{r=1}^{s-1} \binom{i_s - i_r}{j_s - j_r} a_r, \qquad s = 2, \cdots, q,$$

with $a_1 = \begin{pmatrix} i_1 \\ j_1 \end{pmatrix}$ and

(6.1.11)
$$P_A = \frac{n!\,m!}{(n+m)!} \sum_{r=1}^{q} \begin{pmatrix} n+m-i_r \\ n-j_r \end{pmatrix} a_r,$$

which give the required solution. This technique is, of course, just the two-sample version of the method for computing the probability of crossing a general boundary described in § 2.5. One could also develop a two-sample version of the method described in (2.4) but we shall not pursue this possibility. These two methods are essentially the "outside method" and the "inside method" presented from a combinatorial standpoint by Hodges [65] in a useful general review of the two-sample problem.

An alternative solution for a pair of general boundaries has been given by Steck [110] in an explicit determinantal form analogous to (2.5.10) for the one-sample case. Let $b = (b_1, \cdots, b_n)$ and $c = (c_1, \cdots, c_n)$ be two increasing sequences of integers such that $i - 1 \le b_i \le c_i \le m + i + 1$ and as before let $r_i, i = 1, \cdots, n$, be the ranks of the first sample in the ordered combined sample. Steck's result is

(6.1.12)
$$\Pr(b_i < r_i < c_i, i = 1, \cdots, n) = \frac{n!\,m!}{(n+m)!} \det(d_{ij}),$$

where

$$d_{ij} = \begin{pmatrix} c_i - b_j + j - i - 1 \\ j - i + 1 \end{pmatrix}_+, \qquad i, j = 1, \cdots, n.$$

Here,

$$\begin{pmatrix} y \\ z \end{pmatrix}_+ = \begin{cases} 0 & \text{if } z \ne 0 \text{ and } y < z \text{ or if } z < 0, \\ \begin{pmatrix} y \\ z \end{pmatrix} & \text{otherwise.} \end{cases}$$

A short proof of this result has been given by Mohanty [88].

Coming to special cases and taking first the case of equal sample sizes, Gnedenko and Korolyuk [62] showed by simple reflection arguments that

(6.1.13)
$$\Pr(D_{nn}^+ > d) = \frac{\begin{pmatrix} 2n \\ n - c \end{pmatrix}}{\begin{pmatrix} 2n \\ n \end{pmatrix}},$$

where $c = -[-nd]$, and

(6.1.14)
$$\Pr(D_{nn} > d) = \left[2\begin{pmatrix} 2n \\ n - c \end{pmatrix} - \begin{pmatrix} 2n \\ n - 2c \end{pmatrix} + \begin{pmatrix} 2n \\ n - 3c \end{pmatrix} \cdots \right] \bigg/ \begin{pmatrix} 2n \\ n \end{pmatrix}.$$

An elementary exposition of these results is given by Fisz [57, § 10.11]; see also Drion [40].

For the case $m = pn$ with p a positive integer Blackman [18] obtained the explicit result

$$\Pr(D_{nm}^+ > d) = \frac{n!\,m!}{(n+m)!} \sum_{j=0}^{[n-a/p]} \frac{a}{(p+1)j+a} \binom{(p+1)j+a}{j}\binom{(p+1)(n-j)-a}{n-j},$$

(6.1.15)

where $a = -[-md]$; see also Korolyuk [76]. This is the two-sample analogue of (2.1.12) to which it reduces when $m \to \infty$ in an appropriate way.

The complete distribution of D_{nm} has been tabulated for all $n, m \leq 30$ by Lauschbach, von Schweinichen and Wetzel [79] and upper-tail percentage points for the same range of n, m have been tabulated by Wetzel, Jöhnk and Naeve [134]. References to further tabulations are given by Steck [110].

6.2. Limiting distributions of Kolmogorov–Smirnov statistics.
Smirnov [107] proved that the limiting distribution of $[nm/(n+m)]^{1/2}D_{nm}^+$ as $n \to \infty$, $m \to \infty$ such that $n/m \to \lambda > 0$ is the same as that of the corresponding single-sample quantity $\sqrt{n}\,D_n^+$, and similarly for $[nm/(n+m)]^{1/2}D_{nm}^-$ and $[nm/(n+m)]^{1/2}D_{nm}$. However, some work of Hodges [65] showed that the limit is approached rather erratically. Kiefer [71] showed that the condition $n/m \to \lambda > 0$ can be dispensed with and extended the treatment to more than two samples. These results also follow from the fact that for $U(0, 1)$ observations $[nm/(n+m)]^{1/2}[F_n(t) - G_m(t)]$, $0 \leq t \leq 1$, converges weakly to tied-down Brownian motion.

6.3. Cramér–von Mises statistics.
The two-sample Cramér–von Mises statistic is

$$(6.3.1) \qquad W_{nm}^2 = \frac{nm}{n+m} \int_{-\infty}^{\infty} (F_n(x) - G_m(x))^2\, d\left(\frac{nF_n(x) + mG_m(x)}{n+m}\right).$$

This was essentially suggested by Lehmann [81] and was further studied by Rosenblatt [99] but both used a definition less convenient than (6.3.1) in which $d[(nF_n(x) + mG_m(x))/(n+m)]$ was replaced by $d[(F_n(x) + G_m(x))/2]$. The definition (6.3.1) was substituted by Kiefer [71] and Fisz [56] and has been used by later workers.

For practical work the alternative form

$$(6.3.2) \qquad W_{nm}^2 = \frac{m}{n+m} \sum_{j=1}^{n} \left(\frac{r_j - j}{m} - \frac{j - \frac{1}{2}}{n}\right)^2 + \frac{m(2n+m)}{12nm(n+m)}$$

is preferable, where r_j is the rank in the combined sample of the jth observation in the first sample. Since W_{nm}^2 depends only on the ranks it is a distribution-free statistic. Other simple expressions for W_{nm}^2 are available, for example in Anderson [3] and Hájek and Sidák [63]. The form (6.3.2) is due to Burr [23]. Note that as $m \to \infty$ it converges to the corresponding one-sample form (4.1.7.).

We shall derive some simple recursions for the calculation of the small-sample distribution of W_{nm}^2. Let

$$z_j = \frac{r_j - j}{m} - \frac{j - \frac{1}{2}}{n}$$

and let $Z_k = \sum_j^k z_j^2$, where $\sum_j^k$ denotes summation over all j such that $r_j \leq k$. For the binomial process $\{B_n(i)\}$ considered earlier but with an arbitrary probability of success λ, let

$$p_k(j, z) = \Pr(j \ A\text{'s in the first } k \text{ trials and } Z_k = z).$$

We wish to proceed from $p_k(j, z)$ to $p_{k+1}(j + 1, z)$. The event "$j + 1$ A's in the first $k + 1$ trials and $Z_{k+1} = z$" is the union of the two events "$j + 1$ A's in the first k trials, $Z_k = z$ and the $(k + 1)$th trial gives a B" and "j A's in the first k trials, $Z_k = z - [(k - j)/m - (j + \frac{1}{2})/n]^2$ and the $(k + 1)$th trial gives an A." On taking probabilities we obtain

$$p_{k+1}(j + 1, z) = (1 - \lambda)p_k(j + 1, z) + \lambda p_k\left(j, z - \left(\frac{k - j}{m} - \frac{j + \frac{1}{2}}{n}\right)^2\right),$$

that is,

(6.3.3)
$$p_{k+1}(j + 1, z) - p_k(j + 1, z)$$
$$= -\lambda p_k(j + 1, z) + \lambda p_k\left(j, z - \left(\frac{k - j}{m} - \frac{j + \frac{1}{2}}{n}\right)^2\right).$$

On dividing through by λ^{j+1} and putting $q_k(j, z) = \lim_{\lambda \to 0} p_k(j, z)/\lambda^j$, (6.3.3) gives

(6.3.4)
$$\Delta q_k(j + 1, z) = q_k\left(j, z - \left(\frac{k - j}{m} - \frac{j + \frac{1}{2}}{n}\right)^2\right),$$

where Δ is the forward difference operator on k, that is, $\Delta r_k = r_{k+1} - r_k$. The initial conditions for (6.3.4) are $q_0(0, 0) = 1$, $q_0(j, z) = 0$ otherwise.

The required probability is

$$\Pr(W_{nm}^2 = w) = \Pr(Z_n = z | n \ A\text{'s in } n + m \text{ trials})$$

(6.3.5)
$$= \frac{p_n(n, z)}{\binom{n + m}{n} \lambda^n (1 - \lambda)^m}$$

$$= \frac{n! m!}{(n + m)!} q_n(n, z)$$

on letting $\lambda \to 0$, where

$$z = \frac{n + m}{m} w - \frac{m(2n + m)}{12nm(n + m)}.$$

This is similar to the recursion of Hájek and Sidák [63, Theorem b, §IV.3.2], but is simpler to the extent that (6.3.4) only contains two terms whereas the Hájek–Sidák recursion contains three terms.

6.4. Limiting distribution of W_{nm}^2. This was shown by Rosenblatt [99], under the restrictions $m, n \to \infty$ such that $m/n \to \lambda > 0$, to be the same as that of W_n^2 in the one-sample case (see also Fisz [56]). The restriction was removed by Kiefer [71] who also derived the limiting distributions of related statistics derived from more than two samples.

6.5. Two-sample tests on the circle. The statistic

$$(6.5.1) \qquad\qquad V_{nm} = D_{nm}^+ + D_{nm}^-$$

was proposed by Kuiper [78] for testing the hypothesis that two independent samples of n, m observations come from the same distribution on the circle, but it can, of course, be used also for distributions on the line. In fact Gnedenko [61] had previously considered the statistic for the special case $n = m$ for this latter purpose. Like V_n, the value of V_{nm} does not depend on the choice of origin for the computation of $F_n(t)$ and $G_m(t)$. For practical work one may either substitute for D_{nm}^+ and D_{nm}^- from (6.1.5) and (6.1.6) into (6.5.1) or alternatively use the related form given by Maag and Stephens [83]

$$(6.5.2) \qquad V_{nm} = \frac{1}{nm} \left\{ \max_{i=1,\cdots,n} [(n+m)i - nr_i] + \max_{j=1,\cdots,m} [(n+m)j - ms_j] \right\},$$

where r_i, s_i are the ranks of the ith members of the first and second sample respectively in the combined sample.

In order to find the exact distribution of V_{nm} we proceed as for the one-sample statistic V_n in §5.2. Let the points on the circle corresponding to the first sample be denoted by $A_1, \cdots, A_n$. For any choice of origin P, let the particular A_j at which D_{nm}^- is attained, i.e., at which $\max [(r_j - 1)/(n+m) - (j-1)/n]$ is attained, be denoted by A^*. Then as in §5.2 we have

$$\Pr(V_{nm} \leq v) = \Pr(V_{nm} \leq v | P = A^*)$$

$$= \frac{\Pr(V_{nm} \leq v \text{ and } P = A^* | P = A_1 \text{ or } \cdots A_n)}{\Pr(P = A^* | P = A_1 \text{ or } \cdots \text{ or } A_n)}$$

$$(6.5.3) \qquad = n \Pr(V_{nm} \leq v \text{ and } P = A^* | P = A_1 \text{ or } \cdots A_n),$$

where the probability is calculated under a mechanism in which P is chosen from $A_1, \cdots, A_n$ with equal probabilities. As before, this probability is the probability that for a pair of samples of $n-1, m$ observations on the line, the graph of $F_{n-1,m}(i)$ lies between the lines

$$y = -\frac{1}{n-1} + \frac{n}{n-1} \cdot \frac{i}{n+m-1} \quad \text{and} \quad y = \frac{nv-1}{n-1} + \frac{1}{n-1} \cdot \frac{i}{n+m-1},$$

$$(6.5.4) \qquad\qquad\qquad\qquad\qquad\qquad\qquad\qquad i = 1, \cdots, n+m-1.$$

We have therefore reduced the problem on the circle to a problem on the line which has already been solved in § 6.1. By using Steck's determinant (6.1.12), for example, and substituting in (6.5.3) one obtains an explicit expression for the exact distribution function of V_{nm}.

Tables of upper-tail probabilities for various n and m, together with a great deal of information about the exact and asymptotic distributions of V_{nm}, are given by Maag and Stephens [83].

The two-sample statistic of Cramér–von Mises type for tests on the circle is

$$(6.5.5) \quad U_{nm}^2 = \frac{nm}{n+m} \int_{-\infty}^{\infty} \left[F_n(x) - G_m(x) - \int_{-\infty}^{\infty} (F_n(x) - G_m(x)) \, dH(x) \right]^2 dH(x),$$

where

$$H(x) = \frac{nF_n(x) + mG_m(x)}{n+m}.$$

This was introduced by Watson [130]. For practical work the equivalent form

$$(6.5.6) \qquad U_{nm}^2 = \frac{m}{n+m} \sum_{j=1}^{n} (z_j - \bar{z})^2 + \frac{m(m+2n)}{12nm(n+m)}$$

due to Burr [23], where

$$z_j = \frac{r_j - j}{m} - \frac{j - \frac{1}{2}}{n}, \qquad \bar{z} = \frac{1}{n} \sum_{1}^{n} z_j,$$

may be used. A different expression, also suitable for practical calculation, is given by Stephens [115].

The asymptotic distribution of U_{nm}^2 was shown by Watson [130] to be the same as that of U_n^2. Burr [23] tabulates some exact upper-tail probabilities for small values of n, m. Stephens [115] gives the exact first four moments of U_{nm}^2 and uses approximations based on these to extend Burr's tables to larger n, m.

The k-sample statistic of U^2 type has been studied by Maag [82].

7. Tests based on the sample d.f. when parameters are estimated.

7.1. Introductory. So far, all the one-sample tests we have considered have been tests of a fully specified or simple hypothesis. We shall now consider the problem of testing the composite hypothesis $H_0 : F(x) = F_0(x, \theta)$, where $\theta = [\theta_1, \cdots, \theta_p]'$ is a vector of p nuisance parameters whose values are unknown and have to be estimated from the data prior to testing. For example, H_0 might be the hypothesis that the data come from a normal distribution with unknown mean and variance. From a practical standpoint, situations where nuisance parameters are present are much more common than those where they are absent so it is surprising that such a small amount of theoretical work has been done on tests based on the sample d.f. for these situations. Most of this chapter is based on results to appear in [48] for the Kolmogorov–Smirnov tests and [51] for the Cramér–von Mises tests. A technique for testing composite hypotheses by random substitution for unknown parameters suggested in [41] is not recommended because of its lack of power.

The obvious extension of the methods we have considered previously is as follows. Let $\hat{\theta} = [\hat{\theta}_1, \cdots, \hat{\theta}_p]'$ be a suitable estimator of θ, let $\hat{t}_j = F_0(x_j, \hat{\theta})$, $j = 1, \cdots, n$, where $x_1 \leq \cdots \leq x_n$ is the observed sample, and let $\hat{F}_n(t)$ be the sample d.f. calculated from the $\hat{t}_j$'s, that is, $\hat{F}_n(t)$ is the proportion of the values $\hat{t}_1, \cdots, \hat{t}_n$ which are $\leq t$. Now compute D_n^+, D_n, W_n^2, etc. from $\hat{F}_n(t)$. We shall consider the distributions of these statistics beginning with the Kolmogorov–Smirnov statistics for the finite sample case. It will be found that for some cases exact values for the relevant d.f.'s can be obtained under both null and alternative hypotheses.

7.2. Kolmogorov-Smirnov tests, finite-sample case. The only situations of interest to us are those in which the distribution of $\hat{t}_1, \cdots, \hat{t}_n$ does not depend on θ since otherwise the significance points of the test would depend on θ which is unknown. Fortunately, this requirement holds in many cases of practical importance, for example, when θ is a location or scale parameter. For further discussion of this point, see David and Johnson [36] and Darling [33]. We shall assume that this condition is satisfied and also that $\hat{\theta}$ is a complete sufficient statistic. It then follows from a theorem of Basu [8] that $\hat{t}_1, \cdots, \hat{t}_n$ are distributed independently of $\hat{\theta}$. We assume further that $\hat{\theta}$ possesses a density at $\hat{\theta} = \theta_0$ when H_0 is true and $\theta = \theta_0$, where $\theta_0 = [\theta_{01}, \cdots, \theta_{0p}]'$ is any conveniently chosen value of θ.

All the Kolmogorov–Smirnov tests of interest to us can be put in the form: Accept H_0 if $u_j \leq \hat{t}_j \leq v_j, j = 1, \cdots, n$, and reject otherwise. Denoting the acceptance probability when H_0 is true by $\hat{P}$ we therefore have

$$\hat{P} = \Pr\left(u_j \leq \hat{t}_j \leq v_j, j = 1, \cdots, n\right)$$

$$= \Pr\left(u_j \leq \hat{t}_j \leq v_j, j = 1, \cdots, n | \theta_{0i} \leq \hat{\theta}_i \leq \theta_{0i} + d\theta_i, i = 1, \cdots, p\right)$$

for all θ_0 since $\hat{t}_1, \cdots, \hat{t}_n$ are independent of $\hat{\theta}$. Hence,

$$\hat{P} = \frac{\Pr\left(u_j \leq \hat{t}_j \leq v_j, j = 1, \cdots, n \text{ and } \theta_{0i} \leq \hat{\theta}_i \leq \theta_{0i} + d\theta_i, i = 1, \cdots, p | \theta = \theta_0\right)}{\Pr\left(\theta_0 \leq \hat{\theta}_i \leq \theta_{0i} + d\theta_i, i = 1, \cdots, p | \theta = \theta_0\right)}$$

(7.2.1)

$$= \frac{g_0(\theta_0)}{g_u(\theta_0)}$$

on letting $\max_i d\theta_i \to 0$, where $g_0(\theta_0)$ is the density of $\hat{\theta}$ at $\hat{\theta} = \theta_0$ arising from samples satisfying $u_j \leq \hat{t}_j \leq v_j, j = 1, \cdots, n$, and where $g_u(\theta_0)$ is the unrestricted density of $\hat{\theta}$ at $\hat{\theta} = \theta_0$. Both densities are calculated assuming H_0 is true and $\theta = \theta_0$. We assume the existence of $g_0(\theta_0)$ and of the further densities introduced in the following derivation. Since $g_u(\theta_0)$ may be taken to be known we see that the essence of the problem is the calculation of $g_0(\theta_0)$. Our method of solution will be to compute first the Fourier transform of this density and then to obtain $g_0(\theta_0)$ by numerical inversion.

The event "$u_j \leqq \hat{t}_j \leqq v_j, j = 1, \cdots, n$ and $\hat{\theta} = \theta_0$, where $\hat{t}_j = F(x_j, \hat{\theta})$" is identical to the event "$u_j \leqq t_j \leqq v_j, j = 1, \cdots, n$ and $\hat{\theta} = \theta_0$, where $t_j = F(x_j, \theta_0)$." It follows that $g_0(\theta_0)$ can be computed as the density of $\hat{\theta}$ at $\hat{\theta} = \theta_0$ arising from samples satisfying $u_j \leqq t_j \leqq v_j, j = 1, \cdots, n$. This substitution of t_j for $\hat{t}_j$ permits a dramatic simplification since $t_1 \leqq \cdots \leqq t_n$ are uniform order statistics.

We specialize further to estimators of the form

(7.2.2)
$$\hat{\theta} = \frac{1}{n} \sum_{j=1}^{n} h_j(x_j).$$

Though this restriction seems drastic it permits the treatment of important cases such as the estimation of parameters of the exponential and normal distributions. We shall compute the Fourier transform of $g_0(\theta_0)$ by an extension of a method used by Steck [111] to calculate the probability that $u_j \leqq t_j \leqq v_j, j = 1, \cdots, n$.

Let $t'_1, \cdots, t'_n$ denote an *unordered* sample of independent $U(0, 1)$ variables and let

a_i be the event   "$u_i \leqq t'_i \leqq v_i$", $i = 1, \cdots, n,$

b_i be the event   "$t'_{i-1} \leqq t'_i$", $i = 1, \cdots, n$ with $t'_0 = 0,$

c_k be the event   "$u_i \leqq t'_i \leqq v_i, i = 1, \cdots, k$ and $t'_1 \leqq \cdots \leqq t'_k$",

$$k = 1, \cdots, n \text{ with } c_0 \text{ denoting the certain event,}$$

d_{jk} be the event   "$u_k \leqq t'_k < \cdots < t'_{j+1} \leqq v_{j+1}$",

$$j = 0, \cdots, k - 1 \text{   and   } k = 1, \cdots, n.$$

To keep the notation simple let us denote the union of two events α_1, α_2 by $\alpha_1 + \alpha_2$, their difference when $\alpha_1 \supset \alpha_2$ by $\alpha_1 - \alpha_2$ and their intersection by $\alpha_1\alpha_2$. Let b_i^c be the complement of b_i, that is, the event "$t'_{i-1} > t'_i$". Then $a_i = a_ib_i + a_ib_i^c$ so $a_ib_i = a_i - a_ib_i^c$, whence

$$c_k = \prod_{i=1}^{k} a_ib_i = \prod_{i=1}^{k-1} a_ib_i(a_k - a_kb_k^c)$$

$$= a_kc_{k-1} - a_kb_k^c \prod_{i=1}^{k-1} a_ib_i,$$

and so

(7.2.3)
$$c_k = a_kc_{k-1} - a_kb_k^c a_{k-1}c_{k-2} + \cdots + (-1)^{k-j-1} \prod_{i=j+2}^{k} a_ib_i^c a_{j+1}c_j$$

$$+ \cdots + (-1)^{k-1} \prod_{i=2}^{k} a_ib_i^c a_1b_1$$

on repeated substitution of $a_i - a_ib_i^c$ for a_ib_i. The expression (7.2.3) is valid on the understanding that the set operations $-, +, -, \cdots$ are applied in the order indicated proceeding from left to right. This is because each term is a sub-event of the immediately preceding term.

Now $\prod_{i=j+2}^{k} a_i b_i^c a_{j+1}$ is the event "$u_i \leq t_i' \leq v_i$, $i = j + 1, \cdots, k$ and t_{j+1}' $> t_{j+2}' > \cdots > t_k'$" which is the same as the event "$u_k \leq t_k' < \cdots < t_{j+1}' \leq v_{j+1}$", that is, d_{jk} for $j + 2 \leq k$. Substituting in (7.2.3) we obtain

$$(7.2.4) \qquad c_k = \sum_{j=0}^{k-1} (-1)^{k-j-1} c_j d_{jk}, \qquad k = 1, \cdots, n,$$

noting that $d_{k-1,k} = a_k$ for $k = 1, \cdots, n$.

Let $x_1', \cdots, x_n'$ be the values of x corresponding to $t_1', \cdots, t_n'$ when $\theta = \theta_0$, that is, $x_1', \cdots, x_n'$ is an unordered sample of independent observations from the underlying distribution and $t_j' = F(x_j', \theta_0)$ for $j = 1, \cdots, n$. For simplicity let us first consider the case $p = 1$, that is, θ is a scalar parameter. Let $y_j = \sum_{i=1}^{j} h_i(x_i')$, let $\gamma_j(y)$ be the density of y_j arising from paths in c_j, that is,

$$\Pr(c_j \text{ and } w \leq y_j \leq w + dw) = \gamma_j(w) \, dw + o(dw),$$

and let $\delta_{jk}(z)$ be the density of $y_k - y_j$ arising from paths in d_{jk}, that is,

$$\Pr(d_{jk} \text{ and } z \leq y_k - y_j \leq z + dz) = \delta_{jk}(z) \, dz + o(dz).$$

Now for j, k fixed, the events "c_j and $w \leq y_j \leq w + dw$" and "d_{jk} and $z \leq y_k - y_j$ $\leq z + dz$" are independent since the first depends only on $t_1', \cdots, t_j'$ and the second depends only on $t_{j+1}', \cdots, t_k'$ while $t_1', \cdots, t_k'$ are independent random variables. Consequently,

$$\Pr(c_j, d_{jk}, w \leq y_j \leq w + dw \text{ and } z \leq y_k - y_j \leq z + dz)$$
$$= \gamma_j(w) \delta_{jk}(z) \, dw \, dz + o(dw) + o(dz).$$

Putting $y = w + z$ we obtain

$$(7.2.5) \quad \Pr(c_j, d_{jk} \text{ and } y \leq y_k \leq y + dy) = dy \int_{-\infty}^{\infty} \gamma_j(w) \delta_{jk}(y - w) \, dw + o(dy).$$

From (7.2.4) we have by an inclusion-exclusion argument,

$$\Pr(c_k \text{ and } y \leq y_k \leq y + dy) = \sum_{j=0}^{k-1} (-1)^{k-j-1} \Pr(c_j, d_{jk} \text{ and } y \leq y_k \leq y + dy),$$

which on substitution from (7.2.5) and letting $dy \to 0$ gives

$$(7.2.6) \qquad \gamma_k(y) = \sum_{j=0}^{k-1} (-1)^{k-j-1} \int_{-\infty}^{\infty} \gamma_j(w) \delta_{jk}(y - w) \, dw.$$

Let $\gamma_k^*(s)$, $\delta_{jk}^*(s)$ be the Fourier transforms of $\gamma_k(y)$ and $\delta_{jk}(z)$. From (7.2.6) we obtain

$$(7.2.7) \qquad \gamma_k^*(s) = \sum_{j=0}^{k-1} (-1)^{k-j-1} \gamma_j^*(s) \delta_{jk}^*(s), \qquad k = 1, \cdots, n,$$

which is used to compute $\gamma_1^*(s), \cdots, \gamma_n^*(s)$ recursively taking the $\delta_{jk}^*(s)$ as calculable from a knowledge of the underlying distribution of x. Note that $\delta_{jk}(z) = 0$ when $u_k > v_{j+1}$, so $\delta_{jk}^*(s)$ is then zero for all s also.

Alternatively, $\gamma_k^*(s)$ can be expressed explicitly as the determinant

(7.2.8)
$$\gamma_k^* = \begin{vmatrix} \delta_{01}^* & \delta_{02}^* & \cdots\cdots\cdots & \delta_{0k}^* \\ 1 & \delta_{12}^* & & \vdots \\ & & \ddots & \vdots \\ 0 & 1 & \ddots & \\ \vdots & & \ddots & \ddots & \delta_{k-2,k}^* \\ 0 & \cdots\cdots & 0 & 1 & \delta_{k-1,k}^* \end{vmatrix}, \qquad k = 1, \cdots, n,$$

where we have dropped the argument s for convenience and where, as in (7.2.7), $\delta_{jk}^* = 0$ for $u_k > v_{j+1}$.

Taking the inverse transform of $\gamma_n^*(s)$ at $y_n = \theta_0$ we obtain $\gamma_n(\theta_0)$, which by definition satisfies

$$\gamma_n(\theta_0)\, d\theta = \Pr\,(u_i \leq t_i' \leq v_i, i = 1, \cdots, n, t_1' \leq \cdots \leq t_n' \text{ and}$$

$$\theta_0 \leq y_n \leq \theta_0 + d\theta) + o(d\theta).$$

However, our original problem was concerned with the order statistics $t_1 \leq \cdots \leq t_n$, for which $g_0(\theta_0)$ was defined such that

$$g_0(\theta_0)\, d\theta = \Pr\,(u_i \leq t_i \leq v_i, i, \cdots, n \text{ and } \theta_0 \leq \hat\theta \leq \theta_0 + d\theta) + o(d\theta),$$

and this is equivalent to

$$g_0(\theta_0)\, d\theta = \Pr\,(u_i \leq t_i' \leq v_i, i = 1, \cdots, n \text{ and } \theta_0 \leq y_n \leq \theta_0 + d\theta | t_1' \leq \cdots \leq t_n')$$
$$\quad + o(d\theta)$$

$$= \frac{\Pr\,(u_i \leq t_i' \leq v_i, i = 1, \cdots, n, t_1' \leq \cdots \leq t_n' \text{ and } \theta_0 \leq y_n \leq \theta_0 + d\theta)}{\Pr\,(t_1' \leq \cdots \leq t_n')}$$

$$\quad + o(d\theta)$$

$$= n!\gamma_n(\theta_0)\, d\theta.$$

Thus $g_0(\theta) = n!\gamma_n(\theta_0)$ and we obtain finally

(7.2.9)
$$g_0(\theta_0) = \frac{n!}{2\pi} \int_{-\infty}^{\infty} e^{-is\theta_0}\gamma_n^*(s)\, ds,$$

where $\gamma_n^*(s)$ is determined by (7.2.7) or (7.2.8).

Noting the definition of $\gamma_n^*(s)$ as $\int_{-\infty}^{\infty} \exp(isy)\gamma_n(y)\, dy$, we see that $\gamma_v^*(0) = \int_{-\infty}^{\infty} \gamma_n(y)\, dy$ and this is just the probability of the event "$u_i \leq t_i' \leq v_i, i = 1, \cdots,$ n and $t_1' \leq \cdots \leq t_n'$". Thus if $t_1 \leq \cdots \leq t_n$ are the uniform order statistics,

$$\Pr\,(u_i \leq t_i \leq v_i, i = 1, \cdots, n) = n!\gamma_n^*(0).$$

Now

$$\delta_{jk}^*(0) = \Pr\left(u_k \leqq t_k' < \cdots < t_{j+1}' \leqq v_{j+1}\right)$$

$$= \frac{(v_{j+1} - u_k)_+^{k-j}}{(k-j)!}.$$

Substituting in (7.2.8) and taking $k = n$ we obtain Steck's [111] result given in (2.5.10) above.

To illustrate the calculation of δ_{jk}^* consider the exponential distribution for which $F(x, \theta) = 1 - \exp(-x/\theta)$. Take $\theta_0 = 1$, $\hat{\theta} = \bar{x}$ and $y_j = n^{-1}\sum_{i=1}^{j} x_i$. Then $\delta_{jk}^*(s)$ is the Fourier transform of $n^{-1}\sum_{j+1}^{k} x_i$ from samples satisfying $u_k \leqq t_k < \cdots < t_{j+1} \leqq v_{j+1}$, where $t_i = 1 - \exp(-x_i)$. But this is $[(k-j)!]^{-1}$ times the Fourier transform of $n^{-1}\sum_{j+1}^{k} x_i$ from unordered samples satisfying $u_k \leqq t_{j+1}', \cdots, t_k' \leqq v_{j+1}$, that is,

$$\delta_{jk}^*(s) = \frac{1}{(k-j)!}\left[\int_{-\log(1-u_k)}^{-\log(1-v_{j+1})} e^{in^{-1}xs} e^{-x} dx\right]^{k-j}$$

$$= \begin{cases} \dfrac{1}{(k-j)!}\left[\dfrac{(1-u_k)^{1-in^{-1}s} - (1-v_{j+1})^{1-in^{-1}s}}{1-in^{-1}s}\right]^{k-j} & \text{for } u_k < v_{j+1}, \\[2mm] 0 \quad \text{for} \quad u_k \geqq v_{j+1}. \end{cases}$$

Of course, the calculations have to be done in complex arithmetic but this does not cause any serious difficulties.

We now consider the numerical inversion of $\gamma_n^*(s)$. For a review of a variety of methods, see Bohman [19]. The method favoured by the present author, mainly because of its simplicity, is to use a Fourier series expansion. Taking first the case $\hat{\theta} \geqq 0$, since this holds for the exponential example, one takes the expansion

$$\gamma_n(\theta_0) = \sum_{j=-m}^{m} c_j e^{-2\pi i j \theta_0/b},$$

where the Fourier coefficient c_j is

$$c_j = \frac{1}{b}\int_0^b e^{2\pi i j y/b} \gamma_n(y)\, dy = \frac{1}{b}\gamma_n^*\left(\frac{2\pi j}{b}\right),$$

neglecting the error due to truncation at $y = b$, giving finally

$$(7.2.10) \qquad g_0(\theta_0) = n!\gamma_n(\theta_0) = \frac{n!}{b}\sum_{j=-m}^{m} \gamma^*\left(\frac{2\pi j}{b}\right) e^{-2\pi i j \theta_0/b},$$

where b, m must be taken large enough to give the accuracy required. The accuracy can be investigated by applying the same technique to the determination of the unrestricted density $g_u(\theta_0)$ which is, of course, assumed to be known. For $\hat{\theta}$ distributed about θ_0 the range $(0, b)$ can be replaced by $(\theta_0 - \frac{1}{2}b, \theta_0 + \frac{1}{2}b)$, the result (7.2.10) remaining otherwise unchanged. Similarly for other domains for θ_0.

For the general case $p \geq 1$ we have analogously to (7.2.9) the exact expression

$$(7.2.11) \qquad g_0(\theta_0) = \frac{n!}{(2\pi)^p} \int_{-\infty}^{\infty} \cdots \int_{-\infty}^{\infty} e^{-is'\theta_0} \gamma_n^*(s) \, ds_1 \cdots ds_p,$$

where $s = _\bullet[s_1, \cdots, s_p]'$ and where $\gamma_n^*(s)$ is determined by the multivariate form of (7.2.7) or (7.2.8). Assuming $\hat{\theta}_1, \cdots, \hat{\theta}_p$ to be distributed over the effective ranges $(a_i - \frac{1}{2}b_i, a_i + \frac{1}{2}b_i)$, $i = 1, \cdots, p$, where $a_1, \cdots, a_p$ are given constants and $b_1, \cdots, b_p$ are sufficiently large we may approximate $g_0(\theta_0)$ by the multiple Fourier series

$$(7.2.12) \qquad \begin{aligned} g_0(\theta_0) &= \frac{n!}{b_1 \cdots b_p} \sum_{j_1 = -m_1}^{m_1} \cdots \sum_{j_p = -m_p}^{m_p} \gamma_n^* \left(\frac{2\pi j_1}{b_1}, \cdots, \frac{2\pi j_p}{b_p} \right) \\ &\quad \times \exp\left[-2\pi i \left(\frac{j_1 \theta_{01}}{b_1} + \cdots + \frac{j_p \theta_{0p}}{b_p} \right) \right]. \end{aligned}$$

These results can be used to compute significance points for Kolmogorov–Smirnov and Anderson–Darling statistics and to compute powers.

7.3. Weak convergence of the sample d.f. when parameters are estimated. Taking $\hat{t}_j = F(x_j, \hat{\theta})$ as before, let $\hat{F}_n(t)$ be the estimated sample d.f., that is, the proportion of $\hat{t}_1, \cdots, \hat{t}_n$ which are $\leq t$ and let

$$(7.3.1) \qquad \hat{y}_n(t) = \sqrt{n}\,(\hat{F}_n(t) - t), \qquad\qquad 0 \leq t \leq 1.$$

We wish to consider the weak convergence of the process $\{\hat{y}_n(t)\}$.

Let N be the closure of a neighbourhood of θ_0. Let $g(t, \theta) = \partial F(x, \theta)/\partial\theta$ when this is expressed as a function of t by means of the transformation $t = F(x, \theta)$ and let $g(t) = g(t, \theta_0)$. We assume that the vector function $g(t, \theta)$ is continuous in (θ, t) for all $\theta \in N$ and $0 \leq t \leq 1$. We assume also that $F(x, \theta_0)$ has a density $f(x, \theta_0)$ such that for almost all x the vector of derivatives $\partial \log f(x, \theta_0)/\partial\theta_0$ exists satisfying

$$E\left(\frac{\partial \log f(x, \theta_0)}{\partial\theta_0} \frac{\partial \log f(x, \theta_0)}{\partial\theta_0'} \right) = \mathcal{I},$$

where $\mathcal{I}$ is finite and positive-definite. The estimator $\hat{\theta}$ is assumed to satisfy

$$(7.3.2) \qquad \sqrt{n}(\hat{\theta} - \theta_0) = \frac{1}{\sqrt{n}} \mathcal{I}^{-1} \sum_{j=1}^{n} \frac{\partial \log f(x_j, \theta_0)}{\partial\theta_0} + \varepsilon_n,$$

where $\varepsilon_n \to 0$ in probability. The variance matrix of $\sqrt{n}(\hat{\theta} - \theta_0)$ thus converges to $\mathcal{I}^{-1}$. Under well-known conditions first used for the scalar case by Cramér [31, p. 500], maximum-likelihood estimators have the property (7.3.2).

On these assumptions it is shown in [49] that $\{\hat{y}_n(t)\} \xrightarrow{\mathfrak{D}} \{\hat{y}(t)\}$, where $\{\hat{y}(t)\}$ is a normal process in the metric space (D, d) with mean function zero and co-variance function

$$(7.3.3) \qquad E(y(t)y(t')) = \min(t, t') - tt' - g(t)'\mathcal{I}^{-1}g(t'), \qquad 0 \leq t, t' \leq 1.$$

The implication of this result is that if $h(\hat{y}_n(t))$ is a function of $\hat{y}_n(t)$ which is continuous in metric d, then $h(y_n(t)) \overset{\mathcal{D}}{\rightarrow} h(\hat{y}(t))$. Because of the technicalities involved we shall not attempt to reproduce the proof here. The covariance function (7.3.3) was given by Darling [33] for θ a scalar and by Kac, Kiefer and Wolfowitz [68] for θ consisting of the mean and variance of a normal distribution.

In what follows, stochastic integrals of the form $\int_0^s k(t)\, dw(t)$, where $k(t)$ is square-integrable and $\{w(t)\}$ is the Brownian motion process, are to be interpreted as mean-square limits as in Doob [39, Chap. IX, § 2], and Hájek and Sidák [63, Problems 8, 9 and 10, pp. 239–240]. Let $b = \mathcal{I}^{-1} \int_0^1 \dot{g}(t)\, dw(t)$, where $\dot{g}(t) = dg(t)/dt$ and let $z(t) = w(t) - tw(1) - g(t)'b$. Using the result $\mathcal{I} = \int_0^1 \dot{g}(t)\dot{g}(t)'\, dt$ it is easy to show that $\{z(t)\}$ has covariance function (7.3.3). Since also $z(t)$ is normal with zero mean it has the same distribution as $\hat{y}(t)$. Consider the regression of $w(t)$ on $w(1)$ and b, noting that $w(1)$ and b are uncorrelated since $E(w(1)b) = \mathcal{I}^{-1}\int_0^1 \dot{g}(t)\, dt = 0$. The regression coefficients are t and $g(t)$, respectively. Thus $z(t)$ is equivalent to the residual from regression of $w(t)$ on $w(1)$ and b. By the properties of residuals from normal regression it follows that $z(t)$ is distributed like $w(t)$ conditional on $w(1) = 0$ and $b = 0$. The same can be shown for any finite set of values $z(s_1), \cdots, z(s_k)$. It follows that the process $\{z(t)\}$ can be realized as the Brownian motion process $\{w(t)\}$ subject to $w(1) = 0$ and $b = 0$; since $\hat{y}(t)$ has the same distribution as $z(t)$ the same is true of $\{\hat{y}(t)\}$. Thus the task of finding the limiting distribution of a continuous function of $y_n(t)$, say, $h(y_n(t))$, has been reduced to that of finding the conditional distribution of $h(w(t))$ given $w(1) = 0$ and $b = 0$.

7.4. Limiting distribution of D_n^+. We shall use the result asserted at the end of the last section to study the limiting distribution of D_n^+ when parameters have been estimated. As for the finite-sample case we do this by computing the Fourier transform of a density and inverting this numerically. For simplicity we shall confine ourselves to the case of a scalar parameter, that is, $p = 1$. Let L denote the line $y = \delta$. Then

$$\lim_{n \to \infty} \Pr(\sqrt{n}\, D_n^+ > \delta) = \Pr(\text{a path of } \{z(t)\} \text{ crosses } L)$$

$$= \Pr(\text{a path of } \{w(t)\} \text{ crosses } L | w(1) = 0,\, b = 0)$$

(7.4.1)
$$= \lim_{db \to 0} \frac{\Pr(\text{a path of } \{w(t)\} \text{ crosses } L \text{ and } 0 < b < db | w(1) = 0)}{\Pr(0 \leq b \leq db | w(1) = 0)}$$

$$= \frac{k_L(0)}{k_u(0)},$$

where $k_L(0)$ is the density of b at $b = 0$ arising from paths which have crossed L given $w(1) = 0$ and $k_u(0)$ is the unconditional density of b at $b = 0$. We have the unconditional density here and not the conditional density given $w(1) = 0$ since b and $w(1)$ are independent, a result which follows since they are uncorrelated and normal.

Consider the conditional distribution of $b(t) = \int_0^t \dot{g}(s)\,dw(s)$ given $w(t) = \delta$. We have $E(b(t)^2) = \int_0^t \dot{g}(s)^2\,ds = v(t)$, say. Also, $E(b(t)w(t)) = \int_0^t \dot{g}(s)\,ds = g(t)$ and $E(w(t)^2) = t$. Thus $E(b(t)|w(t) = \delta) = \delta g(t)/t$ and $V(b(t)|w(t) = \delta) = v(t) - g(t)^2/t$, that is, conditional on $w(t) = \delta$, $b(t)$ is $N(\delta g(t)/t, v(t) - g(t)^2/t)$ and so has characteristic function

(7.4.2)
$$E(e^{i\tau b(t)}|w(t) = \delta) = e^{i\tau\delta g(t)/t - \tau^2(v(t) - g(t)^2/t)/2}.$$

In the same way, the characteristic function of $b(t) - b(s)$ given $w(s) = w(t) = \delta$ is

(7.4.3)
$$E(e^{i\tau(b(t) - b(s))}|w(s) = w(t) = \delta) = e^{-\tau^2 v(s,t)/2}, \qquad s < t,$$

where $v(s, t) = v(t) - v(s) - (g(t) - g(s))^2/(t - s)$. Now the probability that a path of $w(t)$ first crosses L in $(s, s + ds)$ given that $w(t) = \delta$ is well known to be $f(s, t)\,ds + o(ds)$, where

(7.4.4)
$$f(s, t) = \frac{\delta}{s}\left[\frac{t}{2\pi s(t - s)}\right]^{1/2} e^{-\delta^2(t - s)/2st}, \qquad s < t.$$

Let the characteristic function of $b(s)$ given that $w(t)$ first crosses L at $t = s$ be $\phi_s(\tau)$. Because of the strong Markovian nature of $\{w(t)\}$, if $w(t)$ first crosses L at s, then conditional on s fixed and $t > s$, $b(s)$ and $b(t) - b(s)$ are independent random variables. Consequently, the conditional characteristic function of $b(t)$ given that the time of first crossing is s and that there is a subsequent crossing at t is $\phi_s(\tau)\exp(-\frac{1}{2}\tau^2 v(s, t))$. Hence on taking all paths satisfying $w(t) = \delta$ and integrating over time of first crossing s we obtain the integral equation

(7.4.5)
$$e^{i\tau\delta g(t)/t - \tau^2(v(t) - g(t)^2/t)/2} = \int_0^t \phi_s(\tau)\, e^{-\tau^2 v(s,t)/2} f(s, t)\,ds, \qquad 0 < t \leq 1,$$

which is to be solved for the unknown $\phi_s(\tau)$. The right-hand side of (7.4.5) when written out in full is

$$\int_0^t \phi_s(\tau)\, e^{-\tau^2[v(t) - v(s) - (g(t) - g(s))^2/(t - s)]/2} \frac{\delta}{s}\left[\frac{t}{2\pi s(t - s)}\right]^{1/2} e^{-\delta^2(1/s - 1/t)/2}\,ds$$

$$= e^{-\tau^2 v(t)/2 + \delta^2/2t}\sqrt{t} \int_0^t \phi_s^*(\tau)\, e^{\tau^2(g(t) - g(s))^2/2(t - s)} \frac{1}{\sqrt{(t - s)}}\,ds,$$

where

(7.4.6)
$$\phi_s^*(\tau) = \phi_s(\tau)\, e^{\tau^2 v(s)/2} s^{-3/2}\, e^{-\delta^2/2s} \frac{\delta}{\sqrt{(2\pi)}}.$$

Substituting in (7.4.5) we obtain

(7.4.7)
$$e^{(i\delta + \tau g(t))^2/2t} = \sqrt{t} \int_0^t \phi_s^*(\tau)\, e^{\tau^2(g(t) - g(s))^2/2(t - s)} \frac{1}{\sqrt{(t - s)}}\,ds, \qquad 0 < t \leq 1.$$

Now $E[(b(1) - b(t))(w(1) - w(t))] = -g(t)$ and $E(w(1) - w(t))^2 = 1 - t$. Thus given $w(t) = \delta$ and $w(1) = 0, b(1) - b(t)$ is $N[\delta g(t)/(1 - t), v(1) - v(t) - g(t)^2/(1 - t)]$. Also, the probability that a path of $\{w(t)\}$ first crosses L in $t, t + dt$ is

$$\frac{\delta}{t} \frac{1}{\sqrt{(2\pi t(1 - t))}} e^{-\delta^2/2t(1 - t)}.$$

Thus the Fourier transform of $k_L(0)$ is

$$\phi(\tau) = \int_0^1 \phi_t(\tau) e^{i\tau\delta g(t)/(1 - t) + \tau^2[v(1) - v(t) - g(t)^2/(1 - t)]/2} \frac{\delta}{t} \frac{1}{\sqrt{(2\pi t(1 - t))}} e^{-\delta^2/2t(1 - t)} dt.$$

On substituting for $\phi_t(\tau)$ from (7.4.6) we obtain

$$(7.4.8) \qquad \phi(\tau) = e^{-\tau^2 v(1)/2} \int_0^1 \phi_t^*(\tau) e^{(i\delta + \tau(t))^2/2(1 - t)} \frac{1}{\sqrt{(1 - t)}} dt.$$

The computational procedure for calculating $k_L(0)$ is then as follows. First solve (7.4.7) numerically, then compute $\phi(\tau)$ from (7.4.8) by numerical integration and finally use the $\phi(\tau)$'s as Fourier coefficients in a Fourier series approximation similar to that of (7.2.10) for the finite-sample case, replacing the b of (7.2.10) by c and taking $\tau = 2\pi j/c$ for $j = -m, -m + 1, \cdots, m$.

This technique can be extended in principle to the case $p > 1$, to the two-boundary case and to curved boundaries, though of course the calculations will become substantially more complicated.

7.5. Cramér–von Mises tests. Since the problem of finding the exact distributions of Cramér–von Mises statistics after parameter estimation seems intractable for the finite-sample case, we shall limit our attention to the asymptotic case. As for the simple-hypothesis case considered in § 4.5 we shall confine ourselves to statistics of the form

$$(7.5.1) \qquad \widehat{W}_n^2 = \int_0^1 \psi(t)\hat{y}_n(t)^2 \, dt,$$

where $\psi(t)$ is continuous and nonnegative for $0 \leq t \leq 1$. Under the conditions of § 7.3 $\widehat{W}_n^2$ converges weakly to

$$(7.5.2) \qquad \widehat{W}^2 = \int_0^1 \psi(t)z(t)^2 \, dt,$$

where $\{z(t)\}$ is the zero-mean normal process with covariance function (7.3.3). We shall consider $\widehat{W}^2$ in relation to $W^2 = \int_0^1 \psi(t)y(t)^2 \, dt$, where $\{y(t)\}$ is the tied-down Brownian motion process.

Let $\xi(t) = \psi(t)^{1/2}y(t)$ and $\hat{\xi}(t) = \psi(t)^{1/2}\hat{y}(t)$. Then $\xi(t)$ has covariance function

$$(7.5.3) \qquad \rho(t, t') = \psi(t)^{1/2}\psi(t')^{1/2}(\min(t, t') - tt')$$

and $\hat{\xi}(t)$ has covariance function

(7.5.4) $\qquad \hat{\rho}(t, t') = \psi(t)^{1/2}\psi(t')^{1/2}(\min(t, t') - tt' - g(t)'\mathscr{I}^{-1}g(t'))$.

Denote the eigenvalues and normalized eigenvectors of $\rho(t, t')$ and $\hat{\rho}(t, t')$ respectively by λ_j, $l_j(t)$ and $\hat{\lambda}_i, \hat{l}_j(t)$, $j = 1, \cdots, n$. Then we have

(7.5.5) $\qquad \displaystyle\int_0^1 \rho(t, t')l_j(t')\,dt' = \lambda_j l_j(t)$

and

(7.5.6) $\qquad \displaystyle\int_0^1 [\rho(t, t') - \psi(t)^{1/2}\psi(t')^{1/2}g(t)'\mathscr{I}^{-1}g(t')]\hat{l}_j(t')\,dt' = \hat{\lambda}_j\hat{l}_j(t)$.

Our objective is to express $\hat{W}^2$ in the Kac–Siegert [67] form $\sum_{j=1}^{\infty} \hat{\lambda}_j z_j^2$, where $\{z_j\}$ is a sequence of independent $N(0, 1)$ variables. In order to use this form in practice the series has to be truncated at say the mth term, where m is large enough to give sufficient accuracy and is small enough to be computationally feasible. Our first task is thus the calculation of the m largest eigenvalues $\hat{\lambda}_j$. We assume λ_j's and $l_j(t)$'s to be known.

Let us assume that $\psi(t)^{1/2}g(t)$ possesses the Fourier series expansion $\psi(t)^{1/2}g(t)$ $= \sum_{j=1}^{\infty} h_j l_j(t)$, where $h_j = \int_0^1 l_j(t)\psi(t)^{1/2}g(t)\,dt$. Multiplying (7.5.6) by $l_i(t)$ and integrating we obtain

$$\lambda_i \int_0^1 l_i(t')\hat{l}_j(t')\,dt' - h_i'\mathscr{I}^{-1}\int_0^1 \psi(t')^{1/2}g(t')\hat{l}_j(t')\,dt' = \hat{\lambda}_j \int_0^1 l_i(t)\hat{l}_j(t)\,dt,$$

that is,

(7.5.7) $\qquad k_{ij}\lambda_i - h_i'\mathscr{I}^{-1}\displaystyle\sum_{r=1}^{\infty} h_r k_{rj} = k_{ij}\hat{\lambda}_j$

on putting $k_{ij} = \int_0^1 l_i(t)\hat{l}_j(t)\,dt$. Let

(7.5.8) $\qquad \mathscr{I}^{-1/2}\displaystyle\sum_{r=1}^{\infty} h_r k_{rj} = c_j$.

Then for $\hat{\lambda}_j \neq \lambda_i$ we have from (7.5.7)

$$k_{ij} = \frac{h_i'\mathscr{I}^{-1/2}c_j}{\lambda_i - \hat{\lambda}_j},$$

whence on substituting in (7.5.8) we obtain

$$\mathscr{I}^{-1/2}\sum_{r=1}^{\infty} \frac{h_r h_r'}{\lambda_r - \hat{\lambda}_j}\mathscr{I}^{-1/2}c_j = c_j.$$

This has a nonzero solution for c_j if and only if

(7.5.9) $\qquad \left| \mathscr{I}^{-1/2}\displaystyle\sum_{r=1}^{\infty} \frac{h_r h_r'}{\lambda_r - \hat{\lambda}_j}\mathscr{I}^{-1/2} - I_p \right| = 0$, where I_p is the $p \times p$ unit matrix. The eigenvalues of $\hat{\rho}(t, t')$ which differ from those of $\rho(t, t')$ are the roots of the determinantal equation (7.5.9). For practical solution the infinite series in (7.5.9) has to be truncated at a point which depends on the speed of convergence of the Fourier series expansion of $\psi(t)^{1/2}g(t)$.

Having approximated $\mathscr{W}^{-2}$ by $\sum_{j=1}^{m} \hat{\lambda}_j z_j^2$ one can then invert the characteristic function of the latter, that is, $\prod_{j=1}^{m} (1 - 2\hat{\lambda}_j i\tau)^{-1/2}$, numerically by the method of Imhof [66] or that of Slepian [104] in order to obtain significance points. For special cases arising in tests of exponentiality and normality Stephens [120] succeeded in finding exact means and variances of $\mathscr{W}^{-2}$. These may be used to improve the accuracy of the approximation in the way described in [51]. Using the above and related methods, significance points for tests on the exponential and normal distributions are tabulated in [51]. Some results on powers against particular alternatives are also given.

The distributions of Cramér–von Mises statistics after parameter estimation have been considered previously by Darling [33], Kac, Kiefer and Wolfowitz [68] and Stephens [120]. Darling studied the case of a single parameter and obtained a variety of results including an expression for the characteristic function but did not proceed to the inversion. The above technique for computing the $\hat{\lambda}_j$'s based on (7.5.9) is closely related to Darling's approach. Kac, Kiefer and Wolfowitz considered the test for normality where both the mean and variance are estimated from the data. They developed a different technique for computing the eigenvalues and used it to calculate the eight largest. They also gave an approximation to the limiting distribution of $\hat{W}_n^2$ based on the four largest eigenvalues. Stephens computed the exact lower moments of the asymptotic distributions of several statistics of Cramér–von Mises type for tests of normality and exponentiality and obtained approximate significance points based on either a Pearson-curve approximation with the correct first four moments or a χ^2 approximation with the correct first three moments.

After effectively completing this monograph two further contributions to the study of tests after parameter estimation came to my attention. In the notation of § 7.3, Tyurin [123] shows essentially that increments $d\hat{y}(t)$ in the $\{\hat{y}(t)\}$ process behave like residuals from regression of increments $dy(t)$ on the p components of $dg(t)$. Suppose that $h(t)$ is another p-dimensional function and $\{z(t)\}$ is the process whose increments $dz(t)$ are residuals from regression of $dy(t)$ on $dh(t)$. Tyurin considers the situation where significance points of a function $g(\hat{y}(t))$ are not available but those of $g(z(t))$ can easily be constructed. He shows how by means of a rotation, $\{\hat{y}(t)\}$ can be transformed to give a process with the same distribution as $\{z(t)\}$. Thus, after transformation, the significance points of $g(z(t))$ can be used as the basis for a test. He applies the technique to construct a test of normality after fitting the mean and variance. The objection to the method is the arbitrariness of $h(t)$. In view of this one would expect that tests so constructed would generally have less power than the corresponding tests based on $\hat{y}(t)$. It is interesting to note that the basic rotation device used is essentially the same as one used by me in § 2 of [46] for a quite different problem concerned with testing least-squares regression residuals for serial correlation.

Rao [97] suggests that for the purpose of constructing tests the process $F_n(x) - F(x, \hat{\theta})$ should be replaced by the process

$$R_n(x) = F_n(x) - F(x, \hat{\theta}) + 2n^{-1} \sum_{j=1}^{n/2} \frac{\partial \log f(x_j, \hat{\theta})'}{\partial \hat{\theta}} \mathcal{J}^{-1} \frac{\partial F(x, \hat{\theta})}{\partial \hat{\theta}}$$

on the ground that $\lim_{n \to \infty} nE(R_n(x)R_n(y)) = F(x, \theta)(1 - F(y, \theta))$ for $x \leqq y$ under usual assumptions on F. Thus tests based on $R_n(x)$ behave asymptotically like the corresponding tests when $F(x, \theta)$ is completely specified. The same is true, though the author does not notice this, for tests based on the process $S_n(x) = F_n(x) - F(x, \hat{\theta}_1)$, where $\hat{\theta}_1$ is a maximum-likelihood estimator based on a randomly-chosen subsample of $n/2$ observations. This follows fairly easily from Rao's result. At first sight this appears to provide a dramatically simple solution to our problem: merely use an estimator based on a half-sample instead of the whole sample and proceed as if θ were known! The snag is that the particular half-sample chosen will usually be found to affect the results of the test to a substantial extent. Thus it is possible for two different investigators to draw substantially different conclusions from the same data. In fact the technique is open to the same objections as the random-substitution device mentioned at the end of the first paragraph of § 7.1 above, to which it is closely related. I am grateful to Dr. Rao for sending me a pre-publication copy of the paper on which abstract [97] was based.

REFERENCES

[1] I. G. ABRAHAMSON, *Exact Bahadur efficiencies for the Kolmogorov–Smirnov and Kuiper one- and two-sample statistics*, Ann. Math. Statist., 38 (1967), pp. 1475–1490.

[2] B. AJNE, *A simple test for uniformity of a circular distribution*, Biometrika, 55 (1968), pp. 343–354.

[3] T. W. ANDERSON, *On the distribution of the two-sample Cramér–von Mises criterion*, Ann. Math. Statist., 33 (1962), pp. 1148–1159.

[4] T. W. ANDERSON AND D. A. DARLING, *Asymptotic theory of certain "goodness of fit" criteria based on stochastic processes*, Ibid., 23 (1952), pp. 193–212.

[5] ———, *A test of goodness of fit*, J. Amer. Statist. Assoc., 49 (1954), pp. 765–769.

[6] R. S. ANDERSSEN, F. R. DE HOOG AND R. WEISS, *On the numerical solution of Brownian motion processes*, (1972), to appear.

[7] D. E. BARTON AND C. L. MALLOWS, *Some aspects of the random sequence*, Ann. Math. Statist., 36 (1965), pp. 236–260.

[8] D. BASU, *On statistics independent of a complete sufficient statistic*, Sankhya, 15 (1955), pp. 377–380.

[9] R. J. BERAN, *Testing for uniformity on a compact homogeneous space*, J. Appl. Probability, 5 (1968), pp. 177–195.

[10] ———, *Asymptotic theory of a class of tests for uniformity of a circular distribution*, Ann. Math. Statist., 40 (1969), pp. 1196–1206.

[11] P. BILLINGSLEY, *Convergence of Probability Measures*, John Wiley, New York, 1968.

[12] Z. W. BIRNBAUM, *Numerical tabulation of the distribution of Kolmogorov's statistic for finite sample size*, J. Amer. Statist. Assoc., 47 (1952), pp. 425–441.

[13] ———, *On the power of a one-sided test of fit for continuous probability functions*, Ann. Math. Statist., 24 (1953), pp. 484–489.

[14] Z. W. BIRNBAUM AND B. P. LIENTZ, *Tables of critical values of some Renyi type statistics for finite sample sizes*, J. Amer. Statist. Assoc., 64 (1969), pp. 870–877.

[15] Z. W. BIRNBAUM AND R. PYKE, *On some distributions related to the statistic* D_n^+, Ann. Math. Statist., 29 (1958), pp. 179–187.

[16] Z. W. BIRNBAUM AND V. K. T. TANG, *Two simple distribution-free tests of goodness of fit*, Rev. Inst. Internat. Statist., 32 (1964), pp. 2–13.

[17] Z. W. BIRNBAUM AND F. H. TINGEY, *One-sided confidence contours for probability distribution functions*, Ann. Math. Statist., 22 (1951), pp. 592–596.

[18] K. BLACKMAN, *An extension of the Kolmogorov distribution*, Ibid., 27 (1956), pp. 513–520. See also correction, Ibid., 29 (1958), pp. 318–322.

[19] H. BOHMAN, *Approximate Fourier analysis of distribution functions*, Ark. Mat., 4 (1959), pp. 99–157.

[20] L. BREIMAN, *Probability*, Addison-Wesley, Reading, Mass., 1968.

[21] D. R. BRILLINGER, *An asymptotic representation of the sample distribution function*, Bull. Amer. Math. Soc., 75 (1969), pp. 545–547.

[22] H. D. BRUNK, *On the range of the difference between hypothetical distribution function and Pyke's modified empirical distribution function*, Ann. Math. Statist., 33 (1962), pp. 525–532.

[23] E. J. BURR, *Small-sample distributions of the two-sample Cramér–von Mises' W^2 and Watson's U^2*, Ibid., 35 (1964), pp. 1091–1098.

[24] F. P. CANTELLI, *Sulla determinazione empirica delle leggi di probabilita*, Giorn. Ist. Ital. Attuari., 4 (1933), pp. 421–424.

[25] J. CAPON, *On the asymptotic efficiency of the Kolmogorov–Smirnov test*, J. Amer. Statist. Assoc., 60 (1965), pp. 843–853.

[26] LI-CHIEN CHANG, *On the exact distribution of the statistics of N.V. Smirnov and their asymptotic expansion*, Shusiua Tsingchang, 1 (1955), pp. 775–790.

[27] ———, *On the exact distribution of the statistics of A. N. Kolmogorov and their asymptotic expansion*, Acta. Math. Sinica., 6 (1955), pp. 55–81.

[28] D. G. CHAPMAN, *A comparative study of several one-sided goodness-of-fit tests*, Ann. Math. Statist., 29 (1958), pp. 655–674.

[29] D. R. COX AND P. A. W. LEWIS, *The Statistical Analysis of Series of Events*, Methuen, London, 1966.

[30] H. CRAMÉR, *On the composition of elementary errors. Second paper: Statistical applications*, Skand. Aktuartidskr., 11 (1928), pp. 141–180.

[31] ———, *Mathematical Methods of Statistics*, Princeton University Press, Princeton, N.J., 1946.

[32] H. E. DANIELS, *The statistical theory of the strength of bundles of threads, I*, Proc. Roy. Soc. Ser. A, 183 (1945), pp. 405–435.

[33] D. A. DARLING, *The Cramér–Smirnov test in the parametric case*, Ann. Math. Statist., 26 (1958), pp. 1–20.

[34] ———, *The Kolmogorov–Smirnov, Cramér–von Mises tests*, Ibid., 28 (1957), pp. 823–838.

[35] ———, *On the theorems of Kolmogorov–Smirnov*, Theor. Probability Appl., 5 (1960), pp. 356–361.

[36] F. N. DAVID AND N. L. JOHNSON, *The probability integral transformation when parameters are estimated from the sample*, Biometrika, 35 (1948), pp. 182–190.

[37] A. P. DEMPSTER, *Generalized D_n^+ statistics*, Ann. Math. Statist., 30 (1959), pp. 593–597.

[38] J. L. DOOB, *Heuristic approach to the Kolmogorov–Smirnov theorems*, Ibid., 20 (1949), pp. 393–403.

[39] ———, *Stochastic Processes*, John Wiley, 1953.

[40] E. F. DRION, *Some distribution-free tests for the difference between two empirical cumulative distribution functions*, Ann. Math. Statist., 23 (1952), pp. 563–574.

[41] J. DURBIN, *Some methods of constructing exact tests*, Biometrika, 48 (1961), pp. 41–55. (See also correction, Ibid., 53 (1966), p. 629.)

[42] ———, *The probability that the sample distribution function lies between two parallel straight lines*, Ann. Math. Statist., 39 (1968), pp. 398–411.

[43] ———, *Tests of serial independence based on the cumulated periodogram*, Inst. Internat. Statist., 42 (1959), pp. 1039–1048.

[44] ———, *Tests for serial correlation in regression analysis based on the periodogram of least-squares residuals*, Biometrika, 56 (1969), pp. 1–15.

[45] ———, *Asymptotic distributions of some statistics based on the bivariate sample distribution function*, Nonparametric Techniques in Statistical Inference, M. L. Puri, ed., Cambridge University Press, London, 1970.

[46] ———, *An alternative to the bounds test for testing for serial correlation in least squares regression*, Econometrica, 38 (1970), pp. 422–429.

[47] ———, *Boundary-crossing probabilities for the Brownian motion and Poisson processes and techniques for computing the power of the Kolmogorov–Smirnov test*, J. Appl. Probability, 8 (1971), pp. 431–453.

[48] ———, *Kolmogorov–Smirnov tests when parameters are estimated*, to appear.

[49] ———, *Weak convergence of the sample distribution function when parameters are estimated*, Ann. Statist., (1973), to appear.

[50] J. DURBIN AND M. KNOTT, *Components of Cramér–von Mises statistics I*, J. Roy. Statist. Soc. Ser. B, 34 (1972), pp. 290–307.

[51] J. DURBIN, M. KNOTT AND C. C. TAYLOR, *Components of Cramér–von Mises statistics II*, to appear.

[52] M. DWASS, *On several statistics related to empirical distribution functions*, Ann. Math. Statist., 29 (1958), pp. 188–191.

[53] ———, *The distribution of a generalized D_n^+ statistic*, Ibid., 30 (1959), pp. 1024–1028.

[54] V. A. EPANACHNIKOV, *The significance level and power of the two-sided Kolmogorov test in the case of small sample sizes*, Theor. Probability Appl., 13 (1968), pp. 686–690.

[55] W. FELLER, *An Introduction to Probability Theory and its Applications*, John Wiley, New York, 1968.

[56] M. FISZ, *On a result by M. Rosenblatt concerning the von Mises–Smirnov test*, Ann. Math. Statist., 31 (1960), pp. 427–429.

[57] ———, *Probability Theory and Mathematical Statistics*, John Wiley, New York, 1963.

[58] R. FORTET, *Les fonctions aléatoires du type de Markoff associées à certaines equations linéaires aux dérivées partielles du type parabolique*, J. Math. Pures. Appl., 22 (1943), pp. 177–243.

[59] I. I. GIKHMAN AND A. V. SKOROKHOD, *Introduction to the Theory of Random Processes*, W. B. Saunders, Philadelphia, 1969.

[60] V. GLIVENKO, *Sulla determinazione empirica della legge di probabilita*, Giorn. Ist. Utal. Attuari., 4 (1933), pp. 92–99.

[61] B. GNEDENKO, *Kriterien für die Unveränderlichkeit der Wahrscheinlichkeitsverteilung von zwei unabhängigen Stichprobenreihen*, (Russian/German summary), Math. Nachr., 12 (1954), pp. 29–66.

[62] B. V. GNEDENKO AND V. S. KOROLYUK, *On the maximum discrepancy between two empirical distributions*, Select. Transl. Math. Statist. and Probability, vol. 1, Inst. Math. Statist. and Amer. Math. Soc., Providence, R.I., 1961, pp. 13–16.

[63] J. HÁJEK AND Z. ŠIDÁK, *Theory of Rank Tests*, Academia, Prague, 1967.

[64] A. B. HOADLEY, *On the probability of large divisions of functions of several empirical cdf's*, Ann. Math. Statist., 38 (1967), pp. 380–381.

[65] J. L. HODGES, *The significance probability of the Smirnov two-sample test*, Ark. Mat., 3 (1958), pp. 469–486.

[66] J. P. IMHOF, *Computing the distribution of quadratic forms in normal variables*, Biometrika, 48 (1961), pp. 419–426.

[67] M. KAC AND A. J. F. SIEGERT, *An explicit representation of a stationary Gaussian process*, Ann. Math. Statist., 18 (1947), pp. 438–442.

[68] M. KAC AND J. WOLFOWITZ, *On tests of normality and other tests of goodness of fit based on distance methods*, Ibid., 26 (1955), pp. 189–211.

[69] S. KARLIN, *A First Course in Stochastic Processes*, Academic Press, New York, 1966.

[70] J. H. B. KEMPERMAN, *The passage Problem for a Stationary Markov Chain*, Statistical Research Monographs, vol. 1, University of Chicago Press, Chicago, Illinois, 1961.

[71] J. KIEFER, *K-sample analogues of the Kolmogorov–Smirnov and Cramér–von Mises tests*, Ann. Math. Statist., 30 (1959), pp. 420–447.

[72] J. KLOTZ, *Asymptotic efficiency of the two sample Kolmogorov–Smirnov test*, J. Amer. Statist. Assoc., 62 (1967), pp. 932–938.

[73] M. KNOTT, *The small-sample power of one-sided Kolmogorov tests for a shift in location of the normal distribution*, Ibid., 65 (1970), pp. 1384–1391.

[74] A. KOLMOGOROV, *Sulla determinazione empirica di una legge di distribuzione*, Giorn. Ist. Ital. Attuari., 4 (1933), pp. 83–91.

[75] V. S. KOROLYUK, *Asymptotic analysis of the distribution of the maximum deviation in the Bernoulli scheme*, Theor. Probability Appl., 4 (1959), pp. 339–366.

[76] ———, *On the discrepancy of empiric distributions for the case of two independent samples*, Select. Transl. Math. Statist. and Probability, vol. 4, Inst. Math. Statist. and Amer. Math. Soc., Providence, R.I., 1963, pp. 105–121.

[77] N. H. KUIPER, *Alternative proof of a theorem of Birnbaum and Pyke*, Ann. Math. Statist., 30 (1959), pp. 251–252.

[78] ———, *Tests concerning random points on a circle*, Nederl. Akad. Wetensch. Proc. Ser. A, 63 (1960), pp. 38–47.

[79] H. LAUSCHBACH, H. B. VON SCHWEINICHEN AND W. WETZEL, *Tabellen der Verteilungsfunktion zum Zwei-Stichproben-Smirnoff-Kolmogorov-Test*, Physica-Verlag, Würzburg, 1967.

[80] H. A. LAUWERIER, *The asymptotic expansion of the statistical distribution of N. V. Smirnov*, Z. Wahrscheinlichkeitstheorie, 2 (1963), pp. 61–68.

[81] E. L. LEHMANN, *Consistency and unbiasedness of certain nonparametric tests*, Ann. Math. Statist., 22 (1951), pp. 165–179.

[82] U. R. MAAG, *A k-sample analogue of Watson's U^2 statistic*, Biometrika, 53 (1966), pp. 570–583.

[83] U. R. MAAG AND M. A. STEPHENS, *The V_{nm} two-sample test*, Ann. Math. Statist., 39 (1968), pp. 923–935.

[84] A. W. MARSHALL, *The small sample distribution of $n\omega_n^2$*, Ibid., 29 (1958), pp. 307–309.

[85] F. J. MASSEY, Jr., *A note on the estimation of a distribution function by confidence limits*, Ibid., 21 (1950), pp. 116–119.

[86] ———, *A note on the power of a non-parametric test*, Ibid., 21 (1950), pp. 440–443.

[87] ———, *Correction to A note on the power of a non-parametric test*, Ibid., 23 (1952), pp. 637–638.

[88] S. G. MOHANTY, *A short proof of Steck's result on two-sample Smirnov statistics*, Ibid., 42 (1971), pp. 413–414.

[89] J. NEYMAN, *"Smooth test" for goodness of fit*, Skand. Aktuartidskr., 20 (1937), pp. 149–199.

[90] M. NOÉ AND G. VANDEWIELE, *The calculation of distributions of Kolmogorov-Smirnov type statistics including a table of significance points for a particular case*, Ann. Math. Statist., 39 (1968), pp. 233–241.

[91] D. B. OWEN, *Handbook of Statistical Tables*, Addison-Wesley, Reading, Mass., 1962.

[92] E. S. PEARSON AND M. A. STEPHENS, *The goodness-of-fit tests based on W_n^2 and U_n^2*, Biometrika, 49 (1962), pp. 397–402.

[93] R. PYKE, *The supremum and infimum of the Poisson process*, Ann. Math. Statist., 30 (1959), pp. 568–576.

[94] ———, *Spacings*, J. Roy. Statist. Soc. Ser. B, 27 (1965), pp. 395–436.

[95] ———, *Empirical processes*, Tech. Rep. 28, Department of Mathematics, University of Washington, 1971.

[96] D. QUADE, *On the asymptotic power of the one-sample Kolmogorov-Smirnov tests*, Ann. Math. Statist., 36 (1965), pp. 1000–1018.

[97] K. C. RAO, *The Kolmogoroff, Cramér-von Mises, chisquare statistics for goodness-of-fit tests in the parametric case*, Abstract 133–6, Bull. Inst. Math. Statist., 1 (1972), p. 87.

[98] A. RÉNYI, *On the theory of order statistics*, Acta. Math. Acad. Sci. Hungar., 4 (1953), pp. 191–227.

[99] M. ROSENBLATT, *Limit theorems associated with variants of the von Mises statistic*, Ann. Math. Statist., 23 (1952), pp. 617–623.

[100] ———, *Random Processes*, Oxford University Press, New York, 1962.

[101] W. SAHLER, *A survey on distribution-free statistics based on distances between distribution functions*, Metrika, 13 (1968), pp. 149–169.

[102] P. SCHMID, *On the Kolmogorov and Smirnov limit theorems for discontinuous distribution functions*, Ann. Math. Statist., 29 (1958), pp. 1011–1027.

[103] V. SESHADRI, M. CSÖRGÖ AND M. A. STEPHENS, *Tests for the exponential distribution using Kolmogorov-type statistics*, J. Roy. Statist. Soc. Ser. B, 31 (1969), pp. 499–509.

[104] D. SLEPIAN, *Fluctuations of random noise power*, Bell System Tech. J., 37 (1958), pp. 163–184.

[105] N. V. SMIRNOV, *Sur la distribution de ω^2 (Critérium de M. R. v. Mises)*, C.R. Acad. Sci. Paris, 202 (1936), pp. 449–452.

[106] ———, *Sur la distribution de ω^2 (Critérium de M. v. Mises)*, (Russian/French summary), Mat. Sbornik (N.S.), 2 (1937) (44), pp. 973–993.

[107] ———, *Sur les écarts de la courbe de distribution empirique*, (Russian/French summary), Ibid., 6 (1939), (48), pp. 3–26.

[108] ———, *Approximate laws of distribution of random variables from empirical data*, Uspekhi Mat. Nauk, 10 (1941), pp. 179–206. (In Russian.)

[109] C. S. SMITH, *A note on boundary crossing probabilities for the Brownian motion*, J. Appl. Probability, to appear.

[110] STECK, G. P., *The Smirnov two-sample tests as rank tests*, Ann. Math. Statist., 40 (1969), pp. 1449–1466.

[111] ———, *Rectangle probabilities for uniform order statistics and the probability that the empirical distribution function lies between two distribution functions*, Ibid., 42 (1971), pp. 1–11.

[112] M. A. STEPHENS, *The distribution of the goodness-of-fit statistic U_n^2. I*, Biometrika, 50 (1963), pp. 303–313.

[113] ———, *The distribution of the goodness-of-fit statistics U_n^2. II*, Ibid., 51 (1961), pp. 393–397.

[114] ———, *The goodness-of-fit statistic V_N: distribution and significance points*, Ibid., 52 (1965), pp. 309–321.

[115] ———, *Significance points for the two-sample statistic $U_{m,n}^2$*, Ibid., 52 (1965), pp. 661–663.

[116] ———, *Statistics connected with the uniform distribution percentage points and application to testing for randomness of directions*, Ibid., 53 (1966), pp. 235–240.

[117] ———, *A goodness-of-fit statistic for the circle, with some comparisons*, Ibid., 56 (1969), pp. 161–168.

[118] ———, *Results from the relation between two statistics of the Kolmogorov-Smirnov type*, Ann. Math. Statist., 40 (1969), pp. 1833–1837.

[119] ———, *Use of the Kolmogorov-Smirnov, Cramér-von Mises and related statistics without extensive tables*, J. Roy. Statist. Soc. Ser. B, 32 (1970), pp. 115–122.

[120] ———, *Asymptotic results and percentage points for goodness-of-fit statistics with unknown parameters*, Mimeographed report, Department of Statistics, Stanford University, California, 1971.

[121] M. A. STEPHENS AND U. R. MAAG, *Further percentage points for W_n^2*, Biometrika, 55 (1968), pp. 428–430.

[122] G. SUZUKI, *Kolmogorov-Smirnov tests of fit based on some general bounds*, J. Amer. Statist. Assoc., 63 (1968), pp. 919–924.

[123] Y. N. TYRURIN, *On parametric hypotheses testing with non-parametric tests*, Theor. Probability. Appl., 15 (1970), pp. 722–725.

[124] B. VAN DER POL AND H. BREMMER, *Operational Calculus Based on the Two-sided Laplace Integral*, Cambridge University Press, London, 1955.

[125] I. VINCZE, *On Kolmogorov-Smirnov type distribution theorems*, Non-parametric Techniques in Statistical Inference, M. L. Puri, ed., Cambridge University Press, London, 1970.

[126] R. VON MISES, *Wahrscheinlichkeitsrechnung*, Wein, Leipzig, 1931.

[127] A. WALD AND J. WOLFOWITZ, *Confidence limits for continuous distribution functions*, Ann. Math. Statist., 10 (1939), pp. 105–118.

[128] ———, *Note on confidence limits for continuous distribution functions*, Ibid., 12 (1941), pp. 118–119.

[129] G. S. WATSON, *Goodness-of-fit tests on a circle*, Biometrika, 48 (1961), pp. 109–114.

[130] ———, *Goodness-of-fit tests on a circle. II*, Ibid., 49 (1962), pp. 57–63.

[131] ———, *Another test for the uniformity of a circular distribution*, Ibid., 54 (1967), pp. 675–677.

[132] ———, *Some problems in the statistics of directions*, Bull. Inst. Internat. Statist., 42 (1969), pp. 374–385.

[133] R. WEISS AND R. S. ANDERSSEN, *A product integration method for a class of singular first-kind Volterra equations*, Numer. Math., to appear.

[134] W. WETZEL, M. JÖHNK AND P. NAEVE, *Statistische Tabellen*, Walter de Gruyter, Berlin, 1967.

[135] E. T. WHITTAKER AND G. N. WATSON, *A Course of Modern Analysis*, 4th ed., Cambridge University Press, London, 1927.

[136] P. WHITTLE, *Some exact results for one-sided distribution tests of the Kolmogorov-Smirnov type*, Ann. Math. Statist., 32 (1961), pp. 499–505.